TALKING ABOUT HEALTH

A Philosophical Dialogue

VIBS

Volume 57

a volume in
Nordic Value Studies
NVS
Heta Häyry and Matti Häyry, Editors

TALKING ABOUT HEALTH

A Philosophical Dialogue

Lennart Nordenfelt

Rodopi

Amsterdam - Atlanta, GA 1997

Cover design by Chris Kok based on a photograph, ©1984 by Robert Ginsberg, of statuary by Gustav Vigeland in the Frogner Park, Oslo, Norway.

∞ The paper on which this book is printed meets the requirements of "ISO 9706:1994, Information and documentation - Paper for documents - Requirements for permanence".

ISBN: 90-420-0284-0

Printed in The Netherlands

CONTENTS

EDITORIAL FOREWORD

During most of the twentieth century philosophical debates in the Nordic countries — Denmark, Finland, Iceland, Norway, and Sweden — have been dominated by two intellectual traditions. These are the analytical line of thought with its close connection to logical positivism, and the hermeneutic-phenomenological approach which has its roots in Catholic philosophy but which has also at times come close to Marxist ideals. Analytical philosophy is best known for its interest in the questions of language, logic, mind, and science, while hermeneutics and phenomenology have been preoccupied by the problems of interpretation and the constitution of human beings as physical, psychological, and cultural entities.

Values have not been particularly prominent as objects of critical study within these traditions. The predecessors of the phenomenological school lost their firmer axiological convictions somewhere between Schopenhauer and Nietzsche, and logical positivists explicitly held that values, as metaphysical chimeras, simply cannot be discussed in a rational manner. Since the Second World War, it has, however, become increasingly clear that values in morality, social life, politics, and elsewhere need to be scrutinized by philosophers regardless of their ideological persuasion. Although the questions concerning norms and values are complex and sometimes confusing, they can and must be studied with commitment and vigor.

Nordic Value Studies, a special series within the Value Inquiry Book Series edited by Robert Ginsberg, is designed to give Nordic scholars an opportunity to present their reflections concerning values to the English-speaking scholarly community. We, as editors of the special series, believe that the distinctly Nordic way of looking at things which has created some of the most extensive and best balanced welfare states in the world can also produce invaluable contributions to axiological studies in many areas, including moral, social, and political philosophy, feminism, religious studies, values in higher education, health care provision, science and technology, history of ideas, and philosophy of culture.

The first volume in the special series is Lennart Nordenfelt's insightful *Talking about Health*, which is a rare combination of what is best in the analytical and interpretive traditions of the Nordic countries.

Heta Häyry and Matti Häyry
Special Series Editors, Nordic Value Studies
Helsinki
June 1997

GUEST FOREWORD

It is a pleasure to welcome a book which re-introduces into philosophy, and exploits with great skill, the classical Socratic method of dialogue.

In health care, philosophy often finds itself torn between the conflicting demands of rigorous argument and effective communication. Lennart Nordenfelt has reconciled these demands. *Talking about Health* retains all the logical sharpness and careful attention to detail characteristic of his earlier books. At the same time, by casting his ideas in the form of a lively debate, he has achieved a remarkable transparency of exposition.

All the main elements of his important theory of health are covered—values, vital goals, personal agency. But these are now brought vividly to life as the debate unfolds through the practival concerns faced daily by patients and health are professionals. Here is also a new and particularly valuable exploration of the implications of Nordenfelt's ideas for mental health.

Socrates was no ivory-tower academic. He taught in the market place in Athens, using his philosophy as a tool to challenge prejudice, to expose compromise. Nordenfelt has put Socrates' tool in the hands of practitioners in the market place of modern health care.

K.W.M. Fulford
Professor of Philosophy and Mental Health
University of Warwick
Honorary Consultant Psychiatrist
University of Oxford
February, 1997

AUTHOR'S PREFACE

This book is in dialogue form. It attempts to communicate the results of research in another way than the traditional academic one. It is nevertheless a way which is of ancient origin and which during the classical age was used particularly for the communcation of philosophical knowledge. My research is philosophical, and the ideas I wish to communicate have in many cases emerged in the course of some sort of dialogue—either the normal sort of dialogue with a friend or colleague, or an inward dialogue with a fictitious person. Presentation of my work in the form of dialogue was therefore a natural choice.

The book has two purposes. First, it attempts to attract new readers to the philosophy of health, readers who have some acquaintance with philosophy but take a deep interest in matters related to health and health-promotion. Second, it is a response to critics among fellow philosophers and health scientists, who have commented on ideas presented in my works, *On the Nature of Health* (1987, 1995) and *Quality of Life, Health, and Happiness* (1993). As a consequence, the book is in one way elementary. Some basic facts about the philosophical enterprise—as well as a few technical concepts—have had to be explained. However, in another way the book is not elementary. My defense here of a holistic theory of health is as comprehensive, and more sensitive, than my earlier more conventionally formulated attempts.

My ambitions on the literary plane are modest. Far be it from me to imagine that through the writing in the form of dialogue I would come anywhere near the heights of Plato and Cicero! In light of this I have not given the conversations a background or context. They can be imagined as taking place wherever the reader may wish—in front of the fire, say, or during a stroll in the country. For me, wide open spaces have been a fine stimulus to philosophical reflection.

The two people engaged in conversation I have christened I and P, the first being the Inquirer and the second the Philosopher. Not that P is always my alter ego. A few times I have crept under the skin of the Inquirer. The reader should suspect when I have doubts about my own opinions. For the most part, however, I is a person outside myself. Sometimes I has been incarnated in one of the many people I have met as a tutor and as a lecturer on the philosophy of health.

I wish to thank my friends and colleagues Per-Erik Liss, Per-Anders Tengland, and Fredrik Svenaeus for their pertinent and helpful comments on the manuscript of this book. Special thanks must be directed to Malcolm Forbes for his excellent work in translating this text into English.

Lennart Nordenfelt
Linköping
June, 1997

One

ON THE PHILOSOPHY OF HEALTH

The Inquirer (I): I have often wanted to talk to you about questions on the philosophy of health and welfare. I have read some of the things you have written, and while I found them interesting I have also felt confused and even provoked. What you write isn't easy to grasp. You have a fondness for academic jargon, for instance, and sometimes you get caught up in technicalities. But I have found ideas that appeal to me, and I appreciate the seriousness of your desire to probe deeply into the philosophical questions that arise in connection with health and welfare. So I would like us to have some conversations where you can set out in as straightforward a way as possible what you and other philosophers have to say about these questions.

The Philosopher (P): Good idea. Where shall we start?

I: I want to start with a basic question. Why have you as a philosopher chosen to devote yourself to the philosophy of health? We are used to philosophers speaking about knowledge, morals, and perhaps God, but not health. Isn't this a new thing for philosophers to speak about?

P: Let me take your second question first. It is true that health has not been in the foreground in Western philosophy. The other things you mention have been far more prominent. Epistemology, the theory of knowledge, has held pride of place in theoretical philosophy, just as moral philosophy, or ethics, has held pride of place in practical philosophy. Philosophical writings on the subject of human health are few and far between.

At the same time, nothing is far-fetched in a philosopher taking up this subject. If anything, it is a subject which philosophy has up to now unjustifiably neglected. Let me elaborate on this. Occupying an extensive area of modern philosophy is theory of science. Its task is to study the theoretical foundations of all the sciences. It investigates what a scientific theory is and should be, and it investigates and elucidates the central scientific concepts.

Theory of science has undergone a strong development since the end of World War II. Its contribution has been especially substantial with regard to the exact sciences, in the first place physics. It has also made great advances with regard to the cultural sciences. But theory of science has not made comparable progress when it comes to the applied sciences. These include medicine and the technical sciences. Only a few philosophers have devoted attention to the applied sciences, and this only during recent decades. In common with several colleagues from my own generation, I have seen the challenge in devoting attention to the theory of an applied science. My choice fell on the theory of medicine.

I: You imply that the philosophy of health is to be approached by way of theory of science. But health care is more than a question of science, isn't it? It's a question of service to humanity.

P: Yes, and that's of the utmost importance. The philosophy of health must pay the greatest attention to care and treatment. The art of healing and care is worth a study of its own from many points of view. Sociologists and anthropologists realized this quite early. Medical sociology and medical anthropology are already major disciplines which have provided us with much knowledge about healing and care in Western and other cultures. At long last, the philosophers are joining in and studying medicine with the aid of their theoretical tools. They ask such questions as the following: What sort of praxis does medicine involve? What is care? How shall we best characterize medicine and care? Are they forms of praxis which have communication as an essential element? If so, what concepts are central to this communication? Here we have a new subject, a new part of philosophy which we can call praxis theory. Strictly speaking, to speak of it as a new subject is incorrect, because considerable work was done on it by Aristotle in the fourth century BC; for instance, he analyzed different concepts of action. Health is the lodestar of the praxis of medicine and care. So here we have a further reason for studying the philosophy of health.

I: Does praxis theory also have to do with morality, with ethics? What becomes of medical ethics otherwise?

P: When I spoke of praxis theory I was not including ethics. In speaking of this theory I am thinking in the first place of the theoretical study of praxis. I am thinking of definitions and theories of the nature of praxis, and also of analyses of the components of praxis. But you are quite right that an interface does exist between medical ethics and the philosophy of health, and the philosophy does affect the ethics. The medical ethicist needs to understand the nature of health. Since health is the main goal of medicine and care, these must have health always in focus, and medical ethics has to take this into account, either considering it problematically or adopting it as a starting point.

I: Do other reasons exist for devoting oneself to the philosophy of health?

P: I can give you more if you wish, but let me just offer you the following one. Health is a major human value, and in our age it has assumed greater importance than ever. For some people, health has even become the meaning of life. Just a few years ago, a Swedish book came out which indicated this in its very title, which in English is, *Health as the Meaning of Life*. Whatever we may think of that, we have to acknowledge that health is high on the scale of values. Health is a state which is related to a value which has in all ages been highly prized: happiness. Some regard health as close to happiness, perhaps almost identical with it, while others think we should draw a sharp line between the two. But whatever the case may be, we see here a strong reason—let us call it a value-theoretical one—for studying the nature of health. Well, now I have given you several good reasons why a philosopher should devote attention to the philosophy of health. I also have some personal reasons, as will emerge during our discussions.

I: Who are your philosophical forerunners? I mean, others must have gone seriously into this subject before.

P: Yes, of course. As so often when we go deep enough into the sources, we find a lot of predecessors. Some of the greatest and most well-known figures in the history of philosophy have devoted attention to the theory of health and medicine. The impression which the textbooks of philosophy give and which was the starting point for our discussion is therefore misleading. Plato and Aristotle had much to say about medicine and the art of healing. We do not have to spend much time studying Plato's *Republic* in order to see the role that the art of healing played for him as a thought-model. And in the modern era such prominent figures as René Descartes and John Locke have been philosophers of medicine too. Locke was a doctor, and he was a close friend of Thomas Sydenham, one of the greatest theoreticians and practitioners in the history of medicine.

Coming closer to our time, we need to mention Friedrich Nietzsche. Health is one of his main themes, as several modern commentators have noted. In his *Die Fröhliche Wissenschaft* (The Gay Science), we find a characterization of health which presages present-day theories in this field. And if we contemplate late twentieth-century thinking from the angle of health, the Frenchman Michel Foucault springs most readily to mind. Foucault was a philosopher and historian of ideas who often took up medical topics as the point of departure for his reflections. Book titles such as (in English) *The History of Madness* and *The Birth of the Clinic* indicate his interests.

I: But look, you are talking about philosophers now. Haven't any physicians, the people involved in this field in the most practical way, reflected on the nature of what they have been doing?

P: Yes, naturally. Many of the greatest physicians throughout history have also been theoreticians who reflected on the nature of medicine and of health. This is especially the case with regard to Galen. This Greek physician who practiced in Rome performed a gigantic task in documenting the medical knowledge as well as the medical speculation of his time. He is said to have written more than 20,000 pages, though not much of it has been translated into modern languages. Galen played an immense part in giving medical thinking the shape which then remained the norm in Europe right up to the modern era. Several of his basic ideas are as alive today as ever. One idea, for instance, which has come to us from antiquity through Galen is that of health as a balance between the primary parts of the body.

I: Let us leave the history of ideas and speak instead about the modern background of health research. Can you outline some of the background and starting points of your thinking?

P: An obvious point of departure for all health research, including the philosophical, is the general state of health of the world population. Of especial significance is the state of health—dreadful indeed—in the developing countries. It has a lot to tell us from the ethical point of view. It can help us to see what is of high priority and what is not. It can help us to keep our expectations and objectives realistic. What goals can we really set for global health care? What, given different prior conditions, is an acceptable state of health? And within an ethical perspective, can we make a distinction between one culture and another in determining what degree of health is to be regarded as acceptable? The World Health Organization (WHO) still speaks of health for all by the year 2000. The WHO definition of health is at the same time utopian: "Health is a state of complete physical, mental and social well-being and not merely the absence of disease and infirmity." What are we to make of this? Is it meaningful even in rhetorical terms?

Another point of departure for health reserach is our privileged Western state of health. We are now in better health than before, at the same time as our demands concerning care and treatment are greater. Furthermore, the proportion of elderly people, often frail in health and requiring a lot of care, is greater than ever. All in all, even in our privileged Western countries, health care is in a state of crisis regarding financial and other resources. What, then, are the realistic goals at our disposal? What should be our view of the health goal? Do alternative goals exist?

I: Up to now you have spoken about the state of health in the Western countries and in the rest of the world. But surely another starting point must be present-day medical science and present-day care and treatment.

P: Medicine has in the past 150 years or so achieved great success and prestige by establishing itself as an empirical biological science. The empirical had been central to medical research throughout the ages, but the new thing in the nineteenth century was the endeavor to make do as far as possible with purely intrascientific biological theories, without the aid of external philosophical platforms. This was how medicine was regarded. Modern theory of science has reasons for questioning this view. This point of departure has been extraordinarily fruitful. It has led to the development of cellular pathology, physiology and bacteriology, indeed, to the entire range of experimental thinking characteristic of modern medical research.

As a consequence, large areas of medicine have been sucessful. Many of the bacterial infectious diseases can be kept in check through medication; some of them have been eradicated. Other infectious diseases have been combated through vaccination. Not least because of the development of anesthesiology, surgery has been able to achieve previously undreamt-of results in the case of most types of disease.

The effect of all this success has been to confirm, indeed cement, the position of medicine as an experimental science. This has significant consequences regarding the view of the physician's role. The physician is the scientist whose task is to diagnose biological changes and biological abnormalities, and to find and apply the most adequate medical (sometimes surgical) therapy to combat these changes and abnormalities. This is largely also true of the physician who treats our mind, the psychiatrist. The Western psychiatrist, with some exceptions, is a biologically oriented physician. She or he takes the mental problem or mental illness as stemming at least in part from a biological abnormality, first and foremost in the brain but also sometimes in the endocrine system. For this reason, the psychiatrist will be apt to choose a form of therapy directed at the biological change.

Modern medicine, in sum, is a collection of biological sciences founded on the belief that medicine's principal task is to combat biological abnormalities, diseases, and infirmities, and to restore in the person being treated the biologically normal state. Modern hospitals are modelled in accordance with this belief. They are built in such a way as to serve as efficiently as possible the biologically oriented therapy. This therapy is nowadays often high-tech and advanced, and this has caused health care costs to soar. The application of this therapy has from time to time been reasonably questioned.

I: Do we have any reason to question the things you describe? Shouldn't medicine have a mainly biological foundation? Must not therapy follow suit? Though therapy can sometimes be applied haphazardly, indiscriminately, surely that's no reason for questioning the whole approach.

P: You react quickly and defensively. I have not so far called anything into question. Nor do I intend to question much biologically oriented medical research as such. I do want to come back to and discuss what proportion of effort and resources should be expended on such research in light of present health care needs. I will be arguing, you see, that health care requires a lot of other research and reflection if it is to function as it should.

I: You must excuse me if in my eagerness I was looking too far ahead. We are still only at the introductory stage, I know. Up to now I have wanted to hear about the background of your work as a philosopher of health. Let us go back to that, shall we? First, I need to grasp what you and your colleagues do.

I do know (partly from personal experience) what is involved in several professions in the health sector—for instance, the doctor's and the nurse's area. I can also pretty easily imagine what medical research—a clinical study, say—involves. At a pinch, I suppose I can imagine what medical sociologists or anthropologists do when they study how people perceive health and medicine. But what does the philosopher do? What tools does the philosopher use and what results can the philosopher obtain? Who can make use of these results, and can a future patient benefit from what you are doing?

P: You ask a lot of questions at once. I will do my best to answer most of them. I think that such questions be asked and answered is imperative. We must try to define our professional role. First, I must emphasize that the majority of philosophers of health occupy no position in health care as such. They are scholars who sometimes belong to a faculty of medicine, and sometimes, as in my case, to a faculty of philosophy. But we do have colleagues, especially on the North American continent, who are actively involved in health care. These are the hospital-based medical ethicists, philosophically trained people who usually also have medical training. They participate in the training of medical staff and in the discussion of acute ethical problems which arise within the hospital. The question can be whether a particular treatment should be given in the case of a seriously ill aged patient, the question may have to do with terminal care, or the question can concern information to be given to a patient and her or his relatives, and so on. This is a professional role which is sure to be developed in European hospitals too.

But the more theoretically oriented philosophy of health—that which is concerned with the basic questions of medicine and care, and that which is concerned with the analysis of fundamental concepts—can scarcely play a direct role. I do hope, that it can play an important indirect role.

I: We can come back to this in future conversations. It is the thing about which I am most curious.

Two

ON THE TOOLS OF PHILOSOPHY

P: You asked in our first conversation about the tools of philosophy. Traditionally the philosopher's tools are thought and language, but perhaps that laconic answer just makes things more obscure than ever. Let us instead concentrate on the philosopher's main task, which I would summarize as the *interpretation of reality*. The interpretation and understanding of phenomena is a task which unites all philosophers, whatever the school to which they may belong and whatever their orientation may be. This is a task which philosophers have in common with all scentists, though with a great difference in emphasis and procedure. The empirical scientist tests hypotheses by experimenting and observing. In exceptional cases she or he will erect more ambitious theories on the basis of the tested hypotheses. The philosopher, however, does not engage in this type of work at all.

The philosopher often starts from the written formulations of scientific advances. She or he attempts to interpret these formulations and to see how they compare with the formulations of closely related theories or hypotheses. The philosopher may, for instance, find in the formulations categories or concepts which have to be interpreted in order to obtain a coherent picture of the reality which is described. Interpretation of categories and concepts is the perfect task for the philosopher.

I: But is the philosopher concerned only with *scientific* advances? Isn't it also a philosophical task to interpret categories and concepts that have to do with everyday life? Surely this has to come into it when we attempt to understand medicine as a praxis.

P: Yes, you are right. The clinical encounter, which is perhaps the very heart of the praxis of health care, has hardly been described at all in the writings concerning biological medicine. In a textbook of, say, internal medicine, you won't find an analysis of what a clinical encounter is. Nor will you find anything to help you get a better grasp of the chief elements of such an encounter, such as interaction, communication, understanding and sympathy.

For the interpretation of the clinical encounter, the discipline of medicine and care needs assistance from other disciplines: sociology, psychology, communication theory and philosophy.

I: But what's the point of that? Wouldn't we manage just as well with what the empirical scientist accomplishes?

P: It depends on what you mean by manage. Mankind managed in the sense of surviving as a species even in the Stone Age. We can manage on many different levels and with different degrees of success. What I maintain here is that philosophy has a major contribution to make. Let me exemplify what I mean on several levels. When it comes to the first cases, the clearest and most practical ones, I think I can convince you at once. In the last case, it will perhaps not be so easy to convince you, because it requires that you be aware of a particular intellectual need in yourself.

Philosophical interpretation, which is often in practice the interpretation of concepts, has a major role to play for the simple reason that in all contexts, especially in scientific ones, must be a clarity and a lack of ambiguity in communication. We must know what we mean when we speak to one another. A research result presented in unclear language can be completely misunderstood and thereby obstruct advance rather than foster it. In the course of my exploration of medical literature I have often had cause to reflect on this.

I: Can you give some examples?

P: Gladly. In 1971 a celebrated reserach project was carried out called The US/UK Diagnostic Project. Its purpose was to investigate the reliability of psychiatric diagnosis, with the main focus on any differences in diagnosis between American and British psychiatrists. Four hundred psychiatrists, 200 American and 200 British, studied eight patients on video. In the case of each patient was several hours of film, showing the person in conversation and doing everyday things. The 400 psychiatrists then diagnosed each of the patients. From the point of view of international psychiatry, the result was catastrophic. On some points the differences of diagnosis were immense. A systematic difference occurred between the American psychiatrists and the British. The Americans were far more likely to diagnose schizophrenia, and the only possible explanation for this was that the American concept of schizophrenia was different from, and considerably more inclusive than, the British. What the American psychiatrists understood by schizophrenia was in part not what the British psychiatrists understood by it.

You can easily see what enormous consequences such conceptual confusion can involve for scientific communication. Imagine, for instance, that a team of British researchers believe they have discovered a revolutionary method of treating schizophrenic patients, which the Americans then also adopt. What risk is in this? Well, the risk of the Americans' employing the method for some categories of patient quite different from those intended by the British researchers. In the extreme case this would be dangerous.

Assume that a new drug has been tested on subjects in a category A and that these people have been found to benefit from the drug and to experience no side-effects. In the next stage, the drug is given not only to persons of type A but also to persons of types B and C. The C-patients experience serious side-effects. Assume that the A-patients correspond to the British schizophrenics. Assume, too, that the A, B, and C-patients correspond to the American schizophrenics. Then the consequences of uncritically transferring the British discovery to America are all too evident.

Let me also exemplify the dangers from the statistics concerning cause of death. In principle, all countries nowadays keep statistics concerning mortality, in that they register not only deaths but also indicate cause of death. This cause is to be indicated with the aid of a nomenclature established by the World Health Organization. The statistics provide valuable information about the global incidence and prevalence of dangerous diseases, and for this reason they are often used in research on morbidity over time. However, they will not be of great help unless the nomenclature is used uniformly. It is not just that the different causes of death have to be designated with the same nomenclature; also necessary is that all who employ the nomenclature understand it in the same way and use the same criteria for establishing cause of death. Otherwise the same sort of problems will arise as in the case of the US/UK Diagnostic Project. The statistics will be misleading or else unenlightening. But additional problems confront us when it comes to the statistics concerning cause of death. Purely medical difficulties exist, for instance. It is often difficult, if not impossible, to make a reasonable diagnosis concerning a dead person that as a doctor you have not been treating. Then comes the question of how we are to understand the concept of cause of death. What sort of cause are we talking about, or should we be talking about? Should the death certificate give the *entire* cause of a person's death? A moment's reflection tells us that's impossible. A person's death is almost always caused by a number of factors working together, and only to a limited extent can a death certificate identify these factors. So what is called the cause of death and goes into the official statistics is only a part of the cause. Naturally this gives rise to weighty problems of interpretation. Do different physicians all over the world have the same way of discriminating among background factors when attempting to establish cause of death?

I: But isn't that regulated in instructions concerning cause of death statistics?

P: Yes, the World Health Organization, WHO, has made an attempt at regulation. For instance, WHO has introduced such concepts as underlying cause of death, intervening cause of death, and immediate cause of death. Worthy attempts have also been made to define these concepts. The underlying cause occupies a special status: in cases where only one cause is indicated, this shall be noted as the underlying cause. All this may be fine, but several studies have shown that we have a lot of problems to solve before we can achieve reasonably reliable international statistics concerning cause of death.

I hope that these examples have convinced you how dependent the scientific dialogue is upon precise definitions and clarity. This need for precise definitions and clarity applies to more than science; the need applies just as much to treatment and care. It is not good if one doctor puts something quite different into a particular diagnosis than does another, and it becomes especially tricky if the first physician happens to refer the patient to the second.

I: I realize the importance of well-defined concepts and categories. But is it really necessary to be a philosopher to see this and act accordingly? Aren't definitions and the investigation of concepts a part of the methodology of every science?

P: This ought to be the case. But in practice little time is devoted to conceptual questions in scientific education. Therefore, it is often left to the philosophers to draw attention to these things.

The task becomes more demanding, and also more philosophical, if the purpose is larger than to attain precision of linguistic usage by means of lucid definitions. I am thinking of when we take a closer look at the logic of usage and the way in which the concepts are connected to one another. We may feel that the disease concepts which WHO calls upon physicians to use—when, for instance, making a diagnosis—are not classified on a uniform basis. Some of them are in the same group because they involve the same anatomical location, others because of the nature of the disease. Infections constitute a separate group irrespective, by and large, of location. The same goes for tumours. There are other principles of division too. All this brings up the question of the fruitfulness of the system of classification. Can we improve things? Here the systematic expert knowledge of the philosopher can prove useful.

I: But now you seem to be looking upon philosophy as a handmaid of science. You have shown how the philosophical work of interpretation can be used for the scientist's or professional's own technical purposes. Which means you have put the philosopher on a par with, for instance, the lab technician or the statistician. This entails assigning the philosopher an important role, but this role can scarcely be the only one for philosophy in this context. The ordinary person would hardly picture the philosopher as just a scrutiniser of language and theory. What more than this constitutes the philosophical role?

P: Several roles are traditionally associated with the philosopher. Especially plain is the role of the moral philosopher. I am thinking here of a moral philosopher who is ready to offer advice and recommendations concerning moral issues. Academic moral philosophers have for a long time forgone this and instead confined themselves to theoretical studies, for instance, concerning the language of morals. But let us retain the interpretation perspective, the view of the philosopher as interpreter of categories and concepts. Let us also retain the link with medicine, the art of healing and care. I think we will be able to find a more independent, and more classically philosophical, role if we draw closer to the more abstract categories and concepts in the sphere of health care. The majority of classical philosophical conceptual analyses have concerned exceedingly abstract concepts. Plato analyzed, for instance, the concepts of knowledge, love, and justice. Aristotle analyzed the fundamental concepts of logic and the concept of causality. Many modern philosophers have followed this path and analyzed the same concepts, with other results.

The distinctive characteristic shared by philosophers in this classical philosophical tradition is the interest in analyzing fundamental and highly abstract concepts. The notion is that the attempt to comprehend these concepts facilitates comprehension of the basic structure of all thought. The most abstract concepts are also the most general ones, the ones that broadly embrace the world. Take, for instance, the concept of causality, which is a central feature of our general world-picture. Inasmuch as we unquestioningly think in terms of cause and effect, we show that we presuppose that events in the world are connected. We think that every event must be due to something, and that every event must lead to something. We are incapable of imagining the existence of a world without such interconnection. To our human perception, a world must contain causes and effects. The observation of this caused the German philosopher Immanuel Kant to assume that the notion of cause and effect is an essential component of cognition. He called it a category of understanding. In Kant's view, there is no cause-and-effect relation "out there" in the world; it derives instead from our own understanding. Relations of cause and effect are a grid that the mind places over the world in order to get a grasp of it.

I: Would it be true to say, then, that the philosopher investigates cognition rather than "reality" itself?

P: Yes, we can describe his investigation in that way. But the philosopher regards the interest in cognition as well-justified. Cognition being our only means for grasping reality, we must attempt to understand it as thoroughly as possible. I can describe the same thing in another way: the philosopher attempts to warn us against a common type of naivety, that characterizes all prescientific thinking and also, remarkably enough, the thinking of many scientists. The naivety consists in the following notion.

Outside human consciousness exists a simple and unequivocal reality, and by way of our senses an exact picture of this unequivocal reality reaches our consciousness; then this picture can be conveyed—by means of an unequivocal language—to others, these receivers understanding exactly and being able to reproduce within themselves the same picture as the sender has wished to convey.

However, a moment's reflection should show that this notion cannot be correct. Even in our sensations, something drastic is happening to the "raw material" of reality. Our sensory organs and our minds immediately shape our conception of this reality. Here our concepts, especially the most abstract ones, play a major role. So by examining these abstract concepts, we learn something central about ourselves. We learn something about the basic structure of the only reality we can reach.

I: That is fascinating. But can such abstract reasoning be in any way applicable to such a practical concern as health care? Have we anything to gain in this respect by studying cognition?

P: Yes, I think we have, even though we are not then examining such a fundamental property of thought as that involving causality. But I think that we are examining something basic in our notion of humanity when we examine our concept of health. We are also examining something basic in our attitude to our fellow-beings when we examine our concepts in the sphere of medicine and care.

I: You think, then, that the philosopher's interpretation of concepts is an investigation of our minds, our mode of understanding the world, and that it thereby enables us to know ourselves better. Philosophy is a path to self-understanding.

P: Precisely. Philosophizing affords us the opportunity of uncovering, up to a point, our true selves, of delineating the structure and limits of our thinking. Not that we can never achieve completion in this. The process of disclosure is itself limited by our the categories of our understanding: we are locked up in a prison from which we can never escape. The realization of this should give rise to self-searching and humility.

Three

ON CONCEPTUAL ANALYSIS

I: I am beginning to understand some of the roles of the philosopher. I have an idea, too, of what the philosopher can accomplish in the health sector. I'll shortly be asking you about that in detail. But first I need to be clear about a few things to do with method. My main question is this: How does it work, this process of interpretation of which you have spoken? I mean, how does the philosopher set about interpreting a concept? And when can we say that an interpretation is correct? Is it at all meaningful to speak of a correct interpretation? One reason for my asking is that philosophy has assumed such enormously different forms. Plato's writings have the style of imaginative literature. Other philosophers produce dry-as-dust writings with an abundance of axioms, theorems, and lemmas. Some express themselves in symbols reminiscent of mathematics. Some of these philosophers are incomprehensible most of the time. But are they all, the way you see it, interpreting the same things and for the same purposes? If so, how can the results be so different?

P: Now you are confronting me with some extremely difficult questions. I do not pretend to be able to offer entirely satisfactory answers. However, the questions are so central that any philosopher worth the name must reflect on them and attempt to offer answers. Let me first say something about the different forms of presentation. To some extent they are expressions of literary techniques. An author uses different techniques for different practical purposes. Writing for this or that specific audience, the philosopher knows that she or he can take some things for granted but not others, knows that she or he has to explain some phenomena but not others. Differences like this do not have to mean that the philosophers are writing about different things or attempting different sorts of interpretation.

Sometimes, however, the different forms of presentation do reflect differences at a deeper level. On occasions the aim is extreme precision. This is often the case when it comes to philosophers belonging to the school which has long been predominant in the English-speaking and Scandinavian countries. In these philosophers' analyses, formal requirements play a large role. Definitions have to satisfy these requirements, and both ambiguity and inconsistency are rigorously excluded. Not for nothing do these philosophers call themselves scientific philosophers. They also like to enlist the aid of the language of symbolic logic. The idea behind these analyses and definitions is that they must be of use in scientific expositions, so the purpose is in part instrumental.

Other philosophers regard their role more in terms of the never-ending pursuit of self-understanding. Technical precision is here of subordinate significance. They no doubt consider that our thinking and analysis are for the most part merely preliminary, and that therefore we cannot have much point in attempting rigorous definition. For this reason, the language of these philosophers tends towards the literary, and as often tentative and probing.

But now to the deep and difficult question. What procedure shall the philosopher adopt for an interpretation or conceptual analysis? And can the philosopher ever say that her or his interpretation is correct? I make a distinction between the more special task of interpreting a particular text or a particular person's use of a word, and the analysis of a general concept. The first task belongs to the field usually called hermeneutics, and is a task falling especially to literary scholars but with which all scholars working in the cultural field have to concern themselves. The analysis of general categories or general concepts, however, is a typically philosophical task. The views I express here concern only this last task.

A basic idea in almost all philosophy is that our language reflects our general concepts, this in such a way that we can discover the nature of the concept by studying how we use a term. If we wish to analyze the *concept* "chair," for instance, a good method is to study how we use the *term* "chair." The way in which the term is used reveals that by a chair we mean a piece of seating furniture which has a back and is intended for just one person. Thereby we already have quite a reasonable analysis.

I: But how do we pin down just how the word is used? Do we study an enormous quantity of writing? Do we listen to an enormous quantity of talk? Do we spend months and months, or even years and years, extracting what we need for our analysis?

P: No, we soon see the hopelessness of such a method. The analysis would take an absurdly long time and be absurdly circumstantial. There would also be a risk of the analysis being misleading. Just think. Among all the writing and all the talk to which we would have recourse if we adopted this empirical method, we should find many uses of the term that with some justice we would call careless and unconsidered. Some of the uses might be jocular. We would also be sure to come across instances of the term's being used by people whose mother tongue is not ours. In short, the list of uses would contain things regarded as incorrect not only by the language expert but by most other people.

I: So the general concept is not reflected in the "incorrect" uses of the term. But does that mean there is always a correct use of a term? If so, who decides what it is?

P: Good questions! Here we are entering a minefield, where different philosophers adopt different positions. I hope I can say something which is neutral enough but which at the same time carries the discussion forward. In many cases there is no single correct use of a term, no way of defining with much precision the "correct" limits of the field of application of a concept. Almost all our ordinary concepts are vague. A classic philosophical example of a vague concept is that of baldness. Our linguistic intuition tells us that a person is bald if she or he hasn't a hair on the head. But the person would be bald even if she or he did have a few hairs, let's say 50. So where do we draw the line? How many hairs can the person have on the head and still be bald? This question has no correct answer. Our ordinary language is not made to meet this demand for precision. Nor, for the most part, do we encounter such a demand. Should such a demand nevertheless arise—should there, for instance, be a need (say, in the hairdressing profession) for a sharp definition of baldness—we would just have to do our best to find a fruitful stipulation. We would just have to decide on a precise technical use of the term "baldness."

Other concepts can be open in another way. The twentieth-century philosopher Ludwig Wittgenstein, who made pioneering comments on language and concepts, sought to illustrate this openness with the aid of the concept "game." He pointed out the differences between games. Some are simple party games which do not have clear rules, or even rules for when the game is over, and some are well-defined games with lucid rules indicating when the game is over (for instance in terms of winning or losing). Some of the last games, for instance bridge and chess, are completely intellectual. At the other end of the scale are athletic games and musical games. Some of these games have characteristics in common, and perhaps every sort of game is related to another sort of game with which it has something in common. However, it is not the case, said Wittgenstein (and for this he has become famous), that all games have some common characteristic on the basis of which they can all be called games. "Game" is therefore, according to Wittgenstein, a good example of a "family concept." He considered that the great majority of ordinary concepts have this openness, but this view has not won general assent.

We can agree with all these comments yet continue to assert that there terms can be used correctly and incorrectly. The vagueness or openness of a concept does not open the door to anything and everything. A man with 7,000 hairs on his head is not bald. A dinner-party isn't a game. If our language did not offer us this degree of clarity it would be unusable. Not that this prevents situations where philosophers or language experts genuinely disagree about what is a correct or incorrect use of a term.

I: All right, I can agree that terms can be used correctly and incorrectly, even though the line can sometimes be hard to draw. But you still haven't answered the question how we ascertain which uses are the correct ones. Do we ask the language experts? Do we see what the dictionary says?

P: Yes, a dictionary can be useful sometimes. The English philosopher of language John Langshaw Austin found the best English dictionaries a big help to deciding about correct and incorrect uses of words. And assuredly a language expert can help in difficult cases. But usually we have no need to consult the expert or the dictionary. You and I have a reasonably good command of the language. We have lived 40 or 50 years in our language environment and have been communicating with people in our language for almost as long. We know what is meant by the words "chair," "person," and "health." If we sit down and reflect on this, we can analyze these terms, these concepts, purely on the basis of our feeling for the language. True, we sometimes go wrong in our analysis, but 99 times out of 100 this is not because we use the term incorrectly but because our reflection upon how we use it is inadequate.

Another thing is that the philosopher does not usually do her or his conceptual analysis in isolation. She or he thinks of a possible definition of the concept in question, writes it down and then tests it on a friend or colleague. This person may find it too inclusive or too narrow, or defective in some other way. If so, the philosopher has to go away and think of a new and better definition. Plato can be said to have set the pattern for conceptual analysis. He wrote his analyses in the form of dialogues. Often one of those taking part in the dialogue was Socrates, who was Plato's mouthpiece; the partner was different from dialogue to dialogue. Plato's procedure is a little insidious. His alter ego Socrates is invariably spared having to do the spadework. The other person proposes the definitions. Socrates' role is that of the acute critic. Socrates finds the defects and compels the other person to alter the definitions.

In the dialogue betwen Socrates and Theaetetus, for instance, Theaetetus opens the dialogue briskly and self-confidently and seeks the nature of knowledge. He quickly makes up his mind that having knowledge is the same as having a belief. Socrates counters by saying that some beliefs are false. Surely, Theaetetus cannot think that such beliefs are also a part of our knowledge? Theaetetus agrees that they have to be excluded. By knowledge we must mean true beliefs. But Socrates will not accept this either. For we all the time harbor beliefs which may be true but which we nevertheless do not usually call knowledge. Socrates forces Theaetetus to reflect further upon the requirements concerning genuine knowledge. In the end, Theaetetus finds that what distinguishes our true beliefs which constitute knowledge from our other true beleifs is the fact of their being well-founded.

In this way the analysis proceeds dialectically. Proposal, criticism. Fresh proposal, fresh criticism. Each proposal can be likened to a scientific theory which is tested and either accepted or rejected. One theory is tested; if it is rejected, the next is tested; and so on. The only difference between the proposals and the theories is that the proposals are tested against a feeling for the language, while the scientific theories are tested against experiments or observations.

I: I have a good mental image of the procedure, but I still feel confused. Are conceptual analysis and linguistic analysis the same thing? Do we just need to examine our own ordinary language? Is it worth so much attention? Don't we sometimes need new and better concepts? Are such concepts of physics as mass, force, and energy really synonymous with their everyday counterparts? And if we have scientific ambitions, must we not always go far beyond everyday language?

P: What you are saying is crucial. Up to now I have indicated only where the consideration of concepts has to start. From the beginning we have our ordinary language and our ordinary sense of what the words mean, and nothing else. But I have not said that we must *finish* there. You chose an example at the cutting edge of scientific discovery: theoretical physics. There the concepts are an enormous distance from ordinary language. But remember that those concepts have been introduced with the aid of precise definitions. We have nothing like that to support us when we start our questioning, when we ask what knowledge is, what justice is, what health is.

Once the analysis is under way, different things can happen, which can be legitimate for their different purposes. We can have different levels of ambition concerning our analysis. We may not be directly interested in detailing the uses of the term "chair," or the term "knowledge," or the term "health." We may be more interested in finding a concept which constitutes the least common denominator between "chair," "*chaise*," and "*Stuhl,*" or between "knowledge," "*connaissance*," and "*Erkenntnis*," and so on. What we seek to characterize is a stylized chair, stylized knowledge, stylized health. We may believe that we will thereby discover a tool which is more internationally valid and therefore more widely applicable.

Also, as in the case of physics, we can have good reasons for changing concepts or even creating completely new ones which are clearer and more efficient for our purposes. However, in such a case, we must explain what we are doing and why. Furthermore, stipulation is subject to restrictions. We are sure to have reason to look at these questions more closely when we come to the characterization of health.

I: But when we stipulate or construct concepts this can scarcely be a process of interpretation any longer, it must be more of a process of creation. We acquire slightly more space in the prison of our minds.

P: Yes, and another thing is that the philosopher here takes on a new role. The philosophical task is no longer just a matter of interpretation, of the clarification of meanings, the philosopher now participates in the *creation* of meanings. A further thing to be noted is that the philosopher normally collaborates with others in this, since a construction or stipulation often has a purpose lying outside the philosopher's sphere of interest. A scientist or specialist may coin a concept, so to speak, for her or his purposes; then the philosopher, trained in the analysis of concepts, can judge whether the new concept is logically coherent and perhaps also help ascertain whether or not it truly serves its purpose.

I: Now I want to press you on a point which appears of urgent interest to your colleagues the sociologists and anthropologists. They sometimes regard philosophical conceptual analysis as too bound up with a particular period and linguistic environment. Such analysis is normally carried out in only one language, the philosopher's own. So what general value has this analysis? Does it say anything more than how a particular language—and thereby a particular conceptual world—functions?

P: I am glad you ask. The criticism is partly justified. Changes in concepts do occur in the course of time. Different languages do also differ in their concepts. All this has to be kept in mind when doing conceptual analysis. However, these differences do not have to mean that it is impossible to perform conceptual analyses which remain valid far beyond the philosopher's own linguistic community. Look at the term "disease" and its counterparts in the Indo-European linguistic world—*maladie*, *Krankheit, sjukdom, bolezn*, and so on. Quite evidently, these terms have much in common. The respective socio-linguistic frameworks are much the same. Diseases—*maladies, Krankheiten, sjukdomar, boleznye*—are phenomena to which attention is paid by similar professional groups in similar types of institutions in the different countries. Furthermore, a non-stop international interchange is concerned with how to treat these terms. On top of that, a vigorous international effort is directed towards achieving standardization, and uniformity of usage with regard to different terms relating to disease. Obviously, a study of how the English term "disease" is used is highly relevant to understanding the French term *maladie*. Not that the analysis will be applicable to the French term in all respects, of course. But no such claim is ever likely to be made.

I: But what prevents other cultures from having concepts of disease quite different from ours? What prevents, say, a culture from existing in India where your discoveries about our use of the term "disease" are totally inapplicable?

P: If no connection exists between our concept of disease and what might be thought the corresponding concept in the Indian culture, then the following question becomes relevant: What reason have we for calling the Indian concept a concept of disease? If no link exists, we can't translate the Indian term by our term "disease." If we have no straightforward term that appears adequate, we have to resort to a more complicated rendering of the Indian concept into our language.

When I was studying Comparative Religion, I once attempted to familiarize myself with, and to understand, a messianic movement in Kenya. In this movement was a figure called *Mugwe*, and my task was to define this figure. I remember how frustrated I became. It was impossible to pin down Mugwe with the aid of any individual category in my language. I tried several aids. First I thought of calling Mugwe high priest. Certainly he had functions which were of a priestly type but so much more also was involved. Then I called him medicine man. Certainly he did things connected with healing, but this was just a sideline. I even I tried county sheriff. But this, too, was inadequate. Mugwe was Mugwe, and that was it. It was a concept previously unknown to me, a concept of a type of universal leader to which no concept in my language corresponded. So the term Mugwe was untranslatable.

I: That's a good illustrative example. What you say about translation is apposite. If X cannot even approximately be translated by Y, we cannot say that X and Y stand for the same concept. But if X *can* be translated by Y, a least common denominator must exist. And it must be possible to delineate this least common denominator. Perhaps that is what a conceptual analysis attempts.

P: That was a concise and fine summary.

I: Is it time to round off this general part of our discussion? I have tried—with some success—to get you to explain what a process of philosophical interpretation involves, especially in the case of conceptual analysis. I now understand better why there can be such differences between one piece of philosophical writing and another. But I am keen to see what effect all this reasoning will have on our view of health, medicine, and care.

Four

ON MEDICAL CONCEPTS

I: I would be glad if we now concentrate on the area of health, because I would like to get a better idea of what a philosopher has to offer there. One thing that immediately occurs to me is connected with our earlier conversation about language and linguistic analysis. Does just *one* ordinary language exist? In our language does only one usage occur of the term "health" or of the term "disease"? The reason I ask is that in the area of health we have the professionals—physician, nurse, physiotherapist, occupational therapist, etc.—and then we have all the rest of us, non-experts, who have an enormous interest in health and who never miss a chance to go on about our own health or ill-health. Surely a difference exists here between the language of the professional and that of the non-professional. What, in that case, is the conceptual analysis intended to capture? Is it the concept of health of the professionals or of the non-professionals?

P: That's an extremely interesting question that is difficult to answer. You are right in thinking that in the area of health differences exist between the language of the professionals and the language which the rest of us use. However, I would not speak of two different ordinary languages. Ordinary language should be the preserve of lay-persons, while the particular language of the professionals should be called *technical* language.

What makes answering your question complicated (it calls for what would be an interesting empirical investigation) is that the language of those working in the area of health is far from being technical through and through. In fact, the physician and the nurse almost all the time use ordinary language, not least when talking to patients. Sometimes their language is manifestly technical, when they are speaking of things which are remote from everyday life. It may be a question of medical methods, or of pharmaceuticals, for instance. Here the relation between ordinary language and technical language is easily seen: the technical is a genuine addition to the ordinary.

The comparison becomes more difficult when we contemplate the phenomena which actually do exist in the everyday world, for example, health, disease, infirmity, handicap, or for that matter care and treatment. When people in the health professions speak of these things, and when lay-persons speak of them, do the same terms mean the same things?

Here again, some situations are clear while others are not. It is concerning the latter that we should carry out a thorough empirical study. Some clear cases are the following. An occupational therapist talks to a colleague about a handicap, and both of them are careful to use the term "handicap" in accordance with the official WHO definition. According to this definition, a distinction is to be made between disabilities and handicaps. A disability is a restriction or lack of ability to perform an activity. A handicap is a disadvantage for an individual that limits or prevents the fulfillment of a role that is normal for that individual. Such a distinction is unknown to the lay-person.

And the other way round. If I am a patient and am asked by a physician, nurse, etc. whether I am feeling all right, am in pain, or am feeling depressed, she or he does *not* use technical language, but the ordinary language of everyday life. Otherwise no communication would occur!

But then we encounter a difficult grey area, where whether the professional person is using technical language is not obvious. We can imagine that the way in which a professional uses a particular term is influnced by her or his professional role and that thus a partially new concept comes into being. And this can happen without the persons concerned having a lucid definition to fall back upon. I think we should bear in mind the existence of such partially technical language when we are doing our analyses. We must also remember that the language of the medical profession can never become completely technical. Care and treatment presuppose contact with lay-persons, and such contact can be achieved only by means of ordinary language.

I: I would be glad if we now consider the medical concepts. Quite a lot of them exist, and we can hardly go into all of them thoroughly. Then comes the question of whether we should call them medical concepts. I know that you have doubts about it.

P: Yes, the term "medical" involves risks here. One such risk has to do with what we have just been speaking about, the question of technical language. "Medicine" designates either a collection of sciences or a collection of professional activities performed by specially trained staff. So the risk exists that in referring to the concepts as "medical," we might think of them as belonging to a technical environment. That is hardly our purpose. A good argument for starting with the ordinary concepts is that they are the original ones, and, for reasons already mentioned—we must always keep these ordinary concepts alive.

I would therefore like to propose the term "health concepts" as a general term not only for health, healthiness, and soundness, but also for the negative concepts illness, sickness, disease, infirmity, defect, deformity, impairment and handicap, and for all their specific variants. Perhaps the term "health concepts" as a general term for all of these is not wholly adequate, but I hope that I have said enough to avoid misunderstanding. We must treat all these concepts as belonging together, with the further inclusion of the activity concepts belonging to the health sector, namely, medical praxis, nursing and other care. They are all interconnected and should therefore be defined in terms of one another.

I: You speak of defining. Are we aiming at definitions in our conceptual analyses? Your Danish colleague Uffe Juul Jensen has criticized the pursuit of definitions by philosophers, and even such a major figure as Karl R. Popper is critical of the ubiquitous demand for definitions. What is your position on this?

P: I don't think that all conceptual analyses should conclude with definitions. In many cases we can obtain the clarity or knowledge we need without complete definitions. Thus far I agree with Jensen and Popper. However, I do think that definitions are a natural final goal in many contexts, in particular in scientific and professional ones. Often the point is to produce a uniform nomenclature and achieve a consensus regarding the use of terms.

I: But how can a philosopher such as Jensen adopt a critical position with regard to definitions? Does he not see that they are essential to meaningful communication?

P: Oh yes, we can be sure he does. But to do him justice, we should consider three cornerstones of his position. First, he has a thesis about how concepts come into being. Our everyday concepts, he says, have not come into being through definitions but through a praxis—through our way of living and our way of communicating about life. Medical concepts, for instance, have emerged from the praxis of health care and from our language relating to this praxis. Never, or almost never, have we stopped to strictly define these concepts. Yet there the concepts are, says Jensen, and for the most part they function efficiently.

Second, Jensen is critical of the results of celebrated attempts to firmly establish our concepts in the sphere of health with the aid of definitions. Here he focuses chiefly on different versions of the concept of the "fault in the machinery" which involves looking upon diseases as defects in a human body, this body to be looked upon in turn as a mechanical organism whose nature has been determined by biological evolution.

I: But *one* definition may be bad without this meaning that *all* definitions are bad.

P: No. I agree with Jensen's criticism of the concept of the "fault in the machinery" and its numerous progeny. I also agree with his view of the emergence of the natural concepts. They have not come into being through definitions. But from this does not follow that it should be impossible to characterise them. The so-called Oxford philosophers, who were especially to the fore in England in the 1950s and 1960s, were renowned for their apposite and subtle descriptions of everyday linguistic praxis.

I: But you can't think that we should stop at current praxis? Definitions, I have now learnt, should not just describe current linguistic usage, but they should also improve that usage.

P: Yes, I think they should. It appears to me that Jensen has a blind faith in existent praxis. He may think that this praxis cannot be wrong. But with my examples from the US/UK Diagnostic Project, I have shown that a praxis may require to be corrected with the aid of reflection. Definitions are tools of such reflection.

I: You mentioned that Jensen's critical position on definitions contains a third cornerstone.

P: Yes. It is a view which is close to the other two. Jensen sees a danger in that an officially propounded definition unwarrantably "freezes" reality. Such a definition will if anything impede the natural conceptual development which occurs in normal praxis. My answer here is like my earlier answers. Definitions can be of different quality. Furthermore, definitions are never to be taken as gospels; they are proposals which we must all the time expose to critical reflection in collaboration with the practitioners concerned. Another thing is that we have always to be alive to the possible need for alternative definitions. More than one concept of health and more than one concept of disease can exist at the same time.

I: With this establishment of positions I think we are ready to start on our main task.

Five

SOME CLASSIC DEFINITIONS OF HEALTH

I: How do you intend to set about your analysis of health? Are you going to have me play the role of Theaetetus and do the spadework, with you then offering critical comments?

P: The idea of doing it that way is tempting. Nevertheless, I intend to start the ball rolling myself by presenting two dominant definitions of health. The intriguing thing is that they are quite different from each other, indeed they are in conflict. We should carefully elucidate them, contemplate their consequences, and attempt to explain why they are so different.

I: I suspect that the WHO definition is one of them.

P: Quite right. Let me repeat it: "Health is a state of complete physical, mental, and social well-being and not merely the absence of disease and infirmity." This is to be contrasted with the straightforward negative definition of health as the absence of disease. A modern philosopher who has embraced this last definition is Christopher Boorse.

The two definitions are illuminating inasmuch as they represent opposite extremes. But the scale of which they are the extreme points is not the scale extending from utopia to realism. The WHO definition is prodigiously utopian, but even the notion of a complete absence of pathology in the body is utopian. It's more than likely that we all of us always bear some infection or defect. No, the scale I am thinking of concerns the relation between the concepts health and disease. WHO sets these concepts completely apart from each other,while Boorse binds them completely together.

I: Do you think the truth must be somewhere in between?

P: Yes, I do. Well, if not the truth, at least the most fruitful concept of health. But let us consider those definitions further. What is the WHO definition saying? A person is in a state of perfect health if she or he possesses perfect well-being in three essential respects. So health is here identified with the best imaginable situation from the welfare point of view. As far as I can see, we can hardly conceive of a better individual state. (A possible exception might be where an external threat exists of which the person is unaware.) Another way of expressing it is to say that health is identified with perfect happiness or, to use a modern term, perfect quality of life.

I: Is anything wrong in that? Isn't perfect happiness a worthy goal for WHO to set up? Isn't it advantageous to know what is our ultimate goal, even though it be rather unrealistic to believe we shall ever attain it?

P: In itself I see nothing wrong with setting up utopian goals: it is valuable to have an ultimate goal to strive toward despite there being no hope of ever reaching it. What I am against is that WHO should call this ultimate goal health. From every point of view WHO would have been better to call it happiness or high quality of life. In identifying health with happiness or high quality of life, WHO has deprived itself of major intellectual tools. Now WHO can no longer contrast happiness with health. It cannot, for reasons of definition, say that X was an extremely happy man despite the frailty of his health. Or the other way round: WHO cannot say that X was in extremely good health but was an extremely unhappy person. I could very well say such a thing, and I would also know what I meant. So I would lack the ability to express a number of major things if I did not distinguish between health and happiness. As you will have realized by now, I am not against sometimes departing from the basic meanings of ordinary language. We can have good reasons for such departures, for making what philosophers call stipulative definitions. But they must be good reasons. In the WHO case I can see none. I am afraid that a risk of considerable confusion occurs instead.

Imagine a physician, or someone concerned with health policy, who sets out to take the WHO definition seriously. How is this person to understand it? Is she or he to interpret it as meaning that health care in all seriousness must involve a concern with a person's family problems, economic crises, choice of occupation, leisure pursuits, etc.? Does WHO think that health care is to be identified with all political and other activity for the benefit of human welfare? It looks that way.

I: Aren't you straining the WHO definition? Is it truly intended as a practical recommendation for global health care?

P: True, we must carefully examine the context in which a definition has arisen and what its purpose is. I don't for a moment believe that those responsible for establishing the WHO constitution (which includes the definition) at the 1946 world congress wished either to impoverish the language or to make health care omnipotent. I am certain that they had the best of intentions, both ideologically and intellectually. Recoiling from the heinous deeds which in the recent past had been carried out in the name of medicine, they sought to lay down in crystal-clear terms that medicine and care shall truly benefit the individual and never descend to wanton biological manipulation. The WHO constitution and the first modern medico-ethical codes came into being in the same historical context. Against this background we would be wrong, I believe, to look upon the WHO definition as resulting from conceptual analysis. But because of the prestige of WHO, this definition has come to be widely quoted in the contexts of medical science and of health policy. For this reason, we must take it seriously and interpret it with care.

I: But what of the classic medical model which Boorse has refined in our day? It does differentiate between health and happiness. What have you against it?

P: Boorse's version of the medical model has two vital components. Though these components often appear together, no necessary connection exists between them. One of them is that health is the same as the absence of disease: people are completely healthy if they have no diseases whatsoever. The other component is that diseases are identical with abnormal physical and mental functions (or with such physical and mental conditions as cause abnormal functioning). In the case of the first component, the concepts health and disease are tightly bound up with each other. Health cannot be affected by anything other than disease, and as soon as a disease is present a deterioration in health occurs. These are not empirical findings. They follow from the conceptual system.

I: But isn't that a perfectly natural starting-point? In ordinary language, health and disease are closely connected with each other.

P: Yes, a conceptual relationship does exist between health and disease. What I wish to submit to critical scrutiny is the powerful notion that the relationship between these concepts is one of contradiction, in the following way: if you have no disease, you are in perfect health; if you are not in perfect health, you have a disease. I can soon convince you that the notion is untenable. I think also that your own feeling for the language will support me in this.

First, an elementary remark. For the strong thesis to have any prospect of appearing reasonable, the term "disease" has to be understood in an extremely broad sense. Note that quite a number of other terms exist which designate "negative" physical and mental states. We speak of injury, defect, deformity, handicap, etc. Naturally people with all these types of condition shall receive health care. A person injured in a road accident is taken care of at a surgical or orthopedic clinic; a number of congenital defects or deformities, for instance, hunchback, can be treated surgically; people with handicaps, congenital or acquired, are now regularly helped within medical rehabilitation programs. I take for granted that you agree that all these types of condition involve a deterioration in the person's health, but normally we would not use the term "disease" to designate any of them. So the strong thesis needs considerable qualification. We have to stipulate that the term "disease" will cover all these other negative physical and mental conditions too.

I: I see the problem. And then the big question is what the negative conditions are. You just gave the start of a list. We can add more things. What about existential crises? Should they be on the list? Do they involve a deterioration in health?

P: Excellent. We do need to acquire a lucid notion of what a "negative state of health" is. And we cannot acquire such a notion unless we have an understanding of that to which the negativity refers.

What is the positive state which is being negated? From this we have to conclude that a positive characterization of health is needed if we are to solve the problem.

I: But can't Boorse, and the medical model, escape from this predicament with the aid of the second component, the one that says that diseases are abnormal physical and mental functions or such conditions as cause abnormal physical and mental functioning? This component implies a definition of health as the completely normal condition.

P: Good, now you have brought us to the enormous question of health and normality. That the two should be intimately related is a tenacious notion deeply rooted in us all. I find remarkable that it should have such a strong position even though it represents nothing clear-cut whatsoever. We hover between two extremely different types of idea when it comes to normality. Since first one, then the other, of these ideas comes into play, we often risk confounding them.

I: I wonder whether normality is such a large subject as to need a conversation on its own.

Six

ON HEALTH AND NORMALITY

I: You said in our last conversation that two types of normality exist and that we risk mixing them up. What two types are they?

P: One idea of the normal is that it is the usual or frequent. A normal phenomenon is a usual phenomenon. A normal physical state is a usual physical state. So the commonest physical state is the healthiest one. Or, to make the idea more precise and reasonable, we can put it the following way. Every organ of the body has a function. The heart pumps blood, the lungs take up oxygen, the liver cleanses the body of poisons. These organs perform their functions with a typical intensity and efficiency. With the aid of statistical methods we can find intervals which are such that 95% of all heart functions, lung functions, and liver functions fall within them. We can call these intervals normal intervals. Interpreting normality as frequency, functions which fall within these normal intervals are healthy functions. A person all of whose organ functions fall within normal intervals is a healthy person.

I: That appears a lucid and reasonable notion. Doesn't scientific medicine largely work in that way? I have heard that both physiologists and clinical chemists draw up normality or reference tables for all organ functions and also for the presence of different substances in the human body. Much of their work consists of then relating the values of individuals to these normal values.

P: Yes, that is a major basis for the work of physiologists and chemists. But it has its snags. I have attended several conferences where clinical chemists complained about the difficulty of finding their reference values by statistical means alone. They are also becoming increasingly unwilling to relate their values to health concepts. Let me mention some of the difficulties that exist. In the first place, we do not encounter just one normal interval for every organ function: strictly speaking, we encounter an infinite number of them. Think of the heart function. Following the norm, we say perhaps that the pulse rate should be between 50 and 70 beats a minute. If we take the pulse of a lot of people we find that perhaps 95% of them have a pulse between 50 and 70. But at the same time we know that, for instance, a person who has just run the 100 meters has a pulse of perhaps 150, and that a person who is greatly excited or alarmed also has a higher pulse. It would go right against the grain to say that the heart function of such a person is now unhealthy. Instead, we say that the person's pulse is normal, given the circumstances. Since the number of possible circumstances is infinite, the number of possible normal intervals is infinite too. You can see that we encounter great practical problems when we attempt to make use of the notion of the normal interval.

I: But doesn't a rather simple solution exist to this? Standardize the background. We have to see to it that function tests are always performed in the same type of situation. The person must be at rest, must have an empty stomach (or must have eaten only what has been prescribed), the person must not be worried about anything, and so on. Furthermore, the test should perhaps be age and sex-related, because presumably differences exist between the organ functions of the young and those of the old, and between men's and women's function.

P: You are becoming quite a sharp biostatistician. The work is going more and more in the direction you mention. From this comes also the increasing talk of reference values instead of normal values. Intervals are here introduced based on reference to a situation described in detail and involving factors of the type you mention: age, sex, rest, food intake, etc. I do not wish to deny the value of this line of investigation—it gives us a great deal of biological knowledge. What I call in question is the notion that this knowledge gives us an immediate answer to the question of what constitutes health. To complicate the picture further, I will give you a few other things to think about. In the late 1940s, Otto Klineberg analyzed a number of physiological investigations where one compared the ordinary organ functions of Chinese people with those of Americans. Kinberg found interesting differences between the two population groups. Even when it came to such basic things as blood pressure and pulse, and to the presence of such substances as potassium and sodium in the body, the values of one group differed significantly from those of the other. The normal Chinese person was different from the normal American. According to the theory of health which we are now discussing, we may say, therefore, that the healthy Chinese person is different from the healthy American.

I: But isn't this just a question of a new variable to be taken into consideration in the background description: a race variable?

P: I knew you were going to say that. However, I have to disapppoint you. It was impossible, you see, to explain the differences just in terms of race. The same differences existed between naturalized Chinese and naturalized Americans. Americans who had lived in China for quite some time had Chinese values, while Chinese who had lived in America for quite some time had American values. The differences, therefore, have to be explained in terms of environment, therefore. The tempo of life in the natural and cultural environment of China has physiological and biochemical consequences, and the more hectic tempo of life in America has other such consequences. We can increase the number of our background variables by adding environmental ones (both natural and cultural), and we can take lifestyle and habits into consideration. At this point, though, we become aware that such inclusions are impractical and ultimately absurd. But let me make plain once and for all why I cannot see that frequencies can be the criterion of health. I will enlist the aid of two examples. Let us imagine that because of a serious accident at a nuclear power station the population of Europe has been exposed to powerful radioactive radiation. This radiation has caused a change in the genetic disposition of a lot of people. The children then born to these people have such deformed feet that it is difficult for them to learn to walk. With our present criteria, we would assuredly speak of a genetic defect. But how is the supporter of the frequency theory to proceed? Imagine that this defect has become dominant and that the majority of the population have it. That makes it normal and thereby also healthy.

Another example. We were speaking just now of the large conceivable variety of background factors that can affect physiological functions. Let us now consider the type of background factor involving stress. People can be subjected to different sorts of stress, and we must be able to draw up frequency lists of normal reactions to these different sorts. People can be subjected to extreme cold, they can be subjected to blows, and they can be subjected to microbe invasions. Let us imagine that we place 100 people on the bank of a river in Africa where a few billion malaria-carrying mosquitoes have just hatched and are frenziedly seeking animal contact. Imagine that all the 100 people get bitten by a lot of the mosquitoes. What is the normal reaction to these bites? Well, we can assume that most of the 100 will react with a strong infection caused by the parasites transfered to the red blood-cells. So an infection is the most frequent reaction in this type of situation. The infection is the normal reaction. From this point of view, the infected person is the healthy person!

I: That's a remarkable consequence. It does look as if the frequency theory of health collapses completely in the face of these examples. You appear right in saying that usualness in itself can scarcely be the mark of health. The usual can sometimes be negative, and then we don't want it. Furthermore, I see that we encounter so-called "normal variants," biologically speaking, which are not quite usual. Take, for instance, our large blood-groups. They all represent different healthy populations, yet only 5% of the population is type AB.

P: That's an essential point. The unusual is not always something negative, nor is the usual always something positive. Also unusual positive things exist. Intuitively you are indicating a vital element of our notion of health which the frequency theory completely misses but to which the WHO definition pays the greatest attention. Health is a positive value-laden concept. For this reason, a definition which involves looking upon health as something negative or value-neutral can scarcely be adequate.

Here I still express myself cautiously, influenced as I am by my role of philosopher. The relationship between concept and value is not always a straightforward matter. Sometimes it is extremely strong: a value is part of the concept itself, or even constitutes the entire concept, as in the case of goodness and beauty. At other times, the relationship is weaker. The concept can presuppose human values or in some other way be associated with human values. I am convinced that health is value-laden at least in the weaker sense. I am not equally convinced, though I do believe it, that health is also a value concept in the stronger sense.

I: Have we now abandoned the discussion of normality? We have taken up only one of the notions of normality, the frequency one. You said there was another one, often mixed up with the first.

P: The other one is based on the "norm" part of the word "normality." The normal biological function or process is that which follows the norm. Such a function or process is as it should be.

I: But what is this? It appears to get us nowhere. To just say that a person is in good health if her or his functions are as they should be—what information does that provide?

P: Not much, I agree. But in one respect it *is* informative. It is the respect that we have just been discussing. To assert that a person is healthy if, and only if, her or his biological functions are as they should be, is to take up the strongest position regarding whether health is a value concept.

I: But don't the advocates of that position indicate how we are to ascertain, and know once and for all, whether a function is healthy or as it should be?

P: Well, no, I would say that what we often encounter is indecisiveness, unfortunately. Sometimes the point of reference of these advocates is the state of the function as found in the majority of the population, which means that they embrace in its purest form the frequency idea about which we have just been speaking and which we have rejected. But sometimes the point of reference is an idea which has to do with the theory of evolution and which has roughly the following content. Biological evolution has produced a number of species, one of which is humankind. This evolution has all the time involved changes, but it has also involved a fixing and preservation of the most viable species and their distinctive features. Humankind has proved to be a viable species. This species has characteristic features when it comes to structure and to function, and we can observe how these features contribute, through a complicated interaction, to the survival of the individual and of the species. Hereby a standard emerges for the structure and function of humanity as a species. This is the standard that the usual biological textbook conveys.

I: So if I understand the matter correctly, the survival of the individual and the survival of the species are seen as essential criteria according to which the biological functions are to be judged. The functions which contribute to the survival of the individual or of the species are healthy, while those that impede survival are not.

P: Yes, such is the theory. But if we look at this idea more closely, difficulties become apparent. For instance, functions can be related to survival to a greater or a lesser extent. Are we then to say that the ones which contribute most to survival are the healthiest? That is hardly in accord with intuition. Furthermore, many so-called sick functions contribute to survival. Even when a person has heart disease, heart function contributes to individual survival.

What we can say in general is that many sick functions limit or lower the individual's capability of survival. The life-sustaining power is less. This I regard as crystal-clear, and this I also regard as a deeply-rooted part of our usual idea of disease.

I: Here you are on the same wavelength as the biomedical people. I had not expected that. Are we, then, to build upon this basic biological idea of health?

P: Why should you think that I have anything against biologists? Certainly we can build upon the basic biological idea which makes the survival function an essential criterion of health. In this context, I would criticize two things: the uncritical use of statistics when the idea is being put to practical use, and the taking of this idea to be the only one requiring consideration when we judge health.

I: Let us take these things one by one. I don't understand why we need to bring in statistics when we have set health in relation to survival. Didn't we abandon statistics when we abandoned the pure form of the frequency interpretation of what are healthy functions and what are not?

P: No, statistics also, mostly for pragmatic reasons, have a part to play here, when we are attempting the difficult task of drawing the line of demarcation between the healthy and the sick. Let me repeat: the case is not so simple that an organ either contributes to the survival of the individual or the species, or does not contribute (or even constitutes a threat). For the most part, we encounter differences of degree. A diseased heart is worse at carrying out its functions, as is a diseased lung or a diseased liver. We have to find a threshold value somewhere on the continuous scale. An organ which in its function cannot attain this threshold value we regard as diseased. How, though, are we to find the threshold value? Theorists who have developed this idea, such as Boorse, have been tempted to say that the majority shall decide: a threshold value should be such that the vast majority (say 95%) of the population will be above it.

I: And you can't accept that idea.

P: No, not if the purpose is that it contribute to the characterization of health and disease. My arguments are basically the same sort as before, and we can use the example of the malaria-carrying mosquitoes again. The malaria infection which afflicts the majority of the people on the bank of the tropical river is the species-typical defense reaction to the extreme circumstance which the mosquito onslaught represents. So this defense reaction, for the preservation of life, ought to be regarded as a sign of health. But our normal feeling is that malaria is a serious disease which calls for treatment.

I: Then you say that the survival idea cannot be the only one that we are to take into consideration.

P: No, it obviously can't be. I regard as unreasonable to say that all our diseases threaten our survival, and I regard as far-fetched to say that all of them even lower our chances of survival. What can reasonably be said is that all diseases in some sense can weaken us, or reduce our ability to act. However, this weakening is often partial and rather mild, with no real threat to survival. Only if we think in terms of the risk of being attacked by a desperado or wild animal, for example, might we say that the sick person usually has a lower chance of survival than the healthy person. In the context of Western civilization, we must be able to speak of human weakness which does not constitute a direct threat to personal survival. So we must complement the survival idea with an idea which takes account of other human weakness.

I: Now I understand your criticism of the idea of normality in the frequency interpretation. And you are not retracting the idea in the normative interpretation. What you are saying is that it is empty. To say that the organ functions of the healthy person are just as they should be, tells us nothing more than that good health is valued highly. The idea needs filling out with substance. Some thinkers have attempted to make survival this substance. You regard that as far from adequate.

P: That is a good summary.

Seven

ON EQUILIBRIUM AND HOMOEOSTASIS

I: I must now press you about an idea which you have never discussed in print: the idea, coming from classical antiquity, of the equilibrium of the body. Many ancient theorists from Hippocrates to Galen maintained that the distinctive characteristic of the healthy body was that its parts balanced one another to produce a harmony. In the more popularized version of this idea, which we know best, the bodily fluids balance one another. The primary bodily fluids were held to be blood, phlegm, black bile, and yellow bile. If an imbalance occurred in that the body contained too much of one fluid or too little of another, a state of disease existed. I understand that this idea of the equilibrium of the body has experienced a renaissance in modern physiology. Can it not play a part, then, in the delineation of normality and health?

P: Well, yes, it's an interesting and, as you indicate, tenacious idea that the healthy, as also the good in general, consists in a balance of different forces. The idea is well worth reflecting on, but we also need to critically examine it. In the first place, we need to know exactly how we are to understand it. The epistemological question is central. How are we to know that an equilibrium exists? How are we to know that blood, phlegm, black bile, and yellow bile are properly balanced? What should be the criterion? It is not enough to just count and measure as the purist frequency theorists do. I am not sure that we can find a good discussion of this in the classical medico-philosophical literature. What we do find is the pragmatic answer: when an equilibrium exists the person feels well and can perform the activities of daily life.

I: But isn't that a splendid answer? Human well-being is the criterion of the existence of a balance between the parts or the fluids of the body.

P: But don't you see what is happening? We have now introduced a completely new factor into the picture: human well-being as an ultimate criterion of health. We have abandoned pure inspection of the body and its parts; we have abandoned the attempt to find anatomical-physiological criteria of the presence of health. And if we did as you suggest, what benefit have we derived from the concept of equilibrium? What has it brought to the discussion? If well-being is the main thing, we should instead (speaking metaphorically) measure the quantities of blood, phlegm, black bile, and yellow bile which we need in order to experience a sense of well-being. The equilibrium idea can hardly be said to contribute anything of use.

I: I am not at all satisfied. I am convinced that a good interpretation of the equilibrium idea is available to us. Perhaps in classical antiquity it was impossible to show convincingly how it functioned, but modern physiology can show it. The word used for it is "homoeostasis." The normal state of the body is one of homoeostasis. And that is by no means an empty notion. You must be familiar enough with Walter Cannon's modern classic, *The Wisdom of the Body,* which describes how the different functions of the body balance one another in the sense that if anything unusual happens in the body a whole series of mechanisms are triggered which take the body back to its original state. If, for instance, the body temperature falls, the surface blood-vessels contract in order to prevent loss of heat and to maintain the body temperature around the vital organs. In the event of injury, a number of well-known repair mechanisms come into operation and restore the body to its uninjured original state. The body has an immense number of such feedback mechanisms, and they have been mapped in great detail. If we say, then, that health is the physical state which the totality of the feedback mechanisms of the body have an innate disposition to maintain, have we not thereby said something substantial without referring to personal well-being?

P: Good. That is a solidly based and comprehensible interpretation of the equilibrium idea. It can also, as you suggest, be realized completely in physiological—or, for that matter, biochemical—terms. That such an equilibrium exists is a vital piece of knowledge about our body, as also about the bodies of all other living organisms. We should also appreciate that the overall state of equilibrium of the body is usually identical with a state where the person has a feeling of health and well-being and can perform the activities of daily life.

I: How nice: we've solved the puzzle! The physiological and biochemical homoeostasis is the same as the healthy state in a human being as well as in an animal or plant.

P: You want quick and straightforward answers. But unfortunately I have to complicate things for you again. You may have noted that I said that the state of equilibrium is usually identical with a state where the person feels healthy. Unfortunately, this is not always so. As you know, we assume for extremely good reasons that the feedback mechanisms of the body are regulated by our genetic code. We can look upon this code as being programmed to ensure the survival of the species and, for a time, the individual. This time appears to be between 80 and 110 years. We can say that the genetic code directs our body to fulfill the program. This presupposes, however, that the genetic code itself is in perfect condition. If the genetic code is damaged, then all the functions which the damaged element governs will be defective. We now know of many typical types of genetic damage which cause life-long defects, sometimes in the form of degenerative diseases. Intensive efforts are being made to identify such diseases. An enormous international project, the HUGO project, exists for the purpose of mapping the entire human genetic material so that we could identify different types of pathological change with maximum clarity. A strong hypothesis nowadays is that most of our serious diseases have a genetic component. This is not to be taken as implying that a specific genetic disposition is enough to cause the person to contract a particular disease: what is meant is that such a disposition can predispose to the disease.

Do you see now at what I am getting? The genetic programs of people are not always intact. In order, therefore, to ascertain what should be regarded as a healthy point of equilibrium in a person's body, it is not enough to study the person's genetic code. The point of equilibrium, the point toward which the feedback mechanisms are working, can in the case of one person be such that the person shows manifest symptoms of disease and does not feel at all well when it has been reached. Here we get little help from the equilibrium. In such a case as this, the physiological and biochemical equilibrium does not guarantee the person's health as we intuitively perceive health.

I: I feel confused and irritated. We must be able to eliminate the cases of defective genetic codes and concentrate on the normal. I mean, the definition of health must be based on the normal genetic code.

P: There we are again. That everlasting normality! But what is the normal genetic code? Which is the person who bears it? Or are we to start talking about what is usual again? Are we to establish the normal genetic code with the aid of frequencies and majorities? Think of the example where the radioactive fallout changes the genetic code of a majority of the population. Are we in that case to refer to the genetic codes of the majority, even though these codes cause their bearers great suffering and serious disability?

I: I dislike these extreme philosophical examples. Hopefully there will not be a nuclear war and radical genetic change in the case of a majority of the population. So for all practical purposes, we can go by the code that the majority of the population have today.

P: You speak of practical purposes. We can do much that is sensible for practical purposes. I believe also that the notion of normality and the counting of frequencies can be employed for practical purposes. In that case we discuss practical tools with which we can work with under the guidance of critical understanding. But we must not let normality, frequency, practicality color the basic definition of health, otherwise we run the risk that the wrong thing will be the ultimate goal of our actions in the sphere of health care. Let's get to the heart of the matter. Is the essential thing that everyone's genetic code should be identical with the code which happens to be the commonest today? Or is it that everyone should feel well, and be able to love and work, as Freud put it?

I: I think I am beginning to see at what you are getting. I am also beginning to see that the time may have come to conclude the discussion of the entirely biologically based concepts of health and disease. You shall soon have a chance to speak about what you wish to replace them with. Let me first summarize what you have said. You started by rejecting WHO's extreme concept of health. You rejected it not so much because it was utopian as because WHO called the utopian goal "health". "Welfare" or "happiness" might have been a more suitable designation. We then examined the more traditional medical model of health and disease. You said that this model usually has two components: the identification of health with the absence of disease, and the identification of health with biological normality (diseases being biologically abnormal states). We scrutinized two principal senses of normality. According to one sense, the normal is the statistically usual. We were quickly able to see, however, that many biologically unusual states exist which we are disposed to call healthy, just as under some circumstances there can be usual states which we would call sick. So usualness cannot in itself provide an answer.

According to the other principal sense of normality, the normal is that which follows a given norm. This is empty of content until we have indicated the norm. A reasonable, and noteworthy, proposal for specification of this norm is to say that a biological function is normal if it contributes to the survival of the organism. You think that this specification is going in the right direction but is at the same time inadequate. Organ functions do not need to directly threaten survival in order to be considered sick. Finally, I wished to test the idea of health as equilibrium. We were agreed that we had to find an adequate criterion of equilibrium, and we found it in the operation of feedback mechanisms in the body. The problem with this concept, however, was that we must presuppose a completely stable—"healthy"—genetic code in order to be able to identify the homoeostasis with human health.

P: That's a good survey! I would just add a few comments to prevent misunderstanding. It may appear as if we have performed a hypercritical and finical analysis of some biological or biostatistical attempts to characterize health and disease. A conclusion to be drawn from this analysis, I think, is that the concept of health and related concepts cannot completely be characterized in biological or biostatistical terms. Health is just as much a psychosocial and anthropological concept, and this is what I hope—with your help—to demonstrate in our further discussion. Naturally, this does not imply any denial of the importance of the biological basis of health. A profound knowledge of human biology is indispensable for understanding many of the conditions for human health, and also for understanding the mechanisms behind much human disease.

Eight

ON THE ORDINARY MEANING OF HEALTH: A LITTLE ABOUT MENTAL HEALTH

I: We meet again in order to acquire a better understanding of the nature of health. Have you a suitable starting-point for today's discussion? We have completely rejected the two definitions with which we began. Neither the WHO definition nor the medical one was to your taste.

P: The best starting-point for our discussion today is no prefabricated definition but our ordinary understanding of health and disease. The fashion is to speak of the phenomenology of everyday life, the philosophical study of the world of everyday experience. That's not a bad starting-point. Let me start by asking you a question. When were you last ill?

I: Christmas. I had a bad dose of flu with quite a temperature.

P: Let us talk about your period of illness. What were the distinctive characteristics of your condition?

I: First I ached all over. I was freezing and felt generally awful. Finally I felt so bad that I had to stop working and go home. I went straight to bed and took my temperature, which turned out to be almost 104 Fahrenheit (40°C). I had difficulty sleeping. The aches and pains went on for several days, and I was feeling more and more listless. I was eating badly, but for a few days I was unusually thirsty. It was difficult to do anything at all the first week. The listlessness went on for three weeks or more. I didn't feel like a real human being until all this was over.

P: Yes, that's enough for an analysis. You are describing several typical phenomena, which we can easily divide into two groups. One group has to do with experiences, sensations, emotions. The other group has to do with inability to do things. An essential link connects the two groups. The aches and pains and feeling miserable are part of the cause of the inability. Another cause is the enormous fatigue which normally accompanies a serious infectious disease.

Yours is a typical example of somatic disease. We might say that the concepts of pain (in a broad sense, including other uncomfortable sensations such as aching and nausea), of fatigue, and of inability to act constitute together an icon of somatic disease.

I: Do you mean, then, that we could define disease as a conglomerate of pain, fatigue, and inability to act?

P: No, steady! The icon is one thing; definitions are something else. But the icon does give us valuable material with the aid of which we can attempt to produce definitions. However, definitions must fulfill many requirements of completeness and stringency, so we must reflect further.

I: Are you thinking, for instance, of the fact that the different types of mental illness offer special problems when it comes to definition?

P: Yes, that's part of it. But I am also thinking of the fact that the somatic field contains so many variants. Furthermore, I am thinking of the great variation in the causes of pain, fatigue, and inability to act. Some of these causes have nothing whatsoever to do with ill-health or disease.

I: How do you mean?

P: Just think of a simple case like your going jogging, quite some distance. Even if you put no great effort into it, you are likely to feel exhausted afterwards, and perhaps have considerable aches and pains. The next couple of hours you are incapable of doing much. So you seem to fulfill the criteria of the disease icon.

I: But that goes by so quickly. A disease should last at least a day or so to be worthy of the name.

P: A disease which lasts only a couple of hours is one thing that we might not care about or have the time to care about, but what on earth makes you think that a disease cannot be brief?

I: I don't know. I suppose I just introduced a practical criterion here in order to remove athletic activities from the pathogenic map.

P: Be careful how you express yourself. The jogging case had nothing to do with sports injuries.

I: But let us get back to the main question and the disease icon. To start with, then, we have to find one or more icons of mental illness and then identify the right sort of causes of these icons in order to find a functioning definition of disease.

P: Yes, that is a reasonable procedure. If we are lucky we can also find a least common denominator for mental and somatic illness, and then the general definition can be simplified.

I: Have you any hopes of finding a lowest common denominator? Don't enormous differences occur? What similarity exists, for instance, between an infection of the upper respiratory passages and an attack of anxiety, or between cataract and schizophrenia? Is not Wittgenstein's view of general concepts and programs for definitions the most suitable thing in this field?

P: I agree with you that large and interesting differences exist. Nor do I think that the icons of the different types of disease have many features in common. My hope, though, is that we can find *one* essential characteristic which is common not just to all diseases but all forms of ill-health.

I: You have awakened my curiosity. But perhaps I should be kept on tenterhooks a bit longer. Could we not begin by looking at a few other icons, ones that have to do with mental illness? I would be glad if we could then also consider the question of the right causal background.

P: What do you think is a typical icon of mental illness?

I: I know I am in deep water here, but I take a classic icon of mental illness to be madness.

P: Yes, that is a classic icon, though it is toned down in modern psychiatry. Yet madness is easy to pin down. What do you yourself take it to mean?

I: The mad person stands outside the prevalent social norms, violates the conventions and even the laws of thought, and neither thinks nor acts rationally.

P: You are outlining the prevalent conception of the mad person. Let us get a grasp of this basic icon. We can first consider how much it in truth has to do with the field of mental illness, then we can see what applicability it has to a subdivision of this field. To begin with, then, let us test its general applicability. Would you say, when you think about it, that the icon "madness" provides us with a good picture of what mental illness is?

I: No. Many conditions now treated by psychiatrists have nothing at all to do with madness. We have the whole sphere of neuroses, which are sometimes quite mild and partial and which are perfectly compatible with people thinking rationally and fulfilling social norms. Nor is a depressed person mad according to prevalent notions.

P: So madness is of limited value as an icon when we wish to capture the totality of mental illnesses. Only a few serious states of mental illness may be described in terms of madness: the schizophrenic states and possibly some of the manic-depressive states. Schizophrenia is by definition a question of a split mind: the person has difficulty in making different realities cohere into a comprehensible whole. Therefore, the schizophrenic's thinking can proceed on different levels at the same time. The surface stream of thought and the thoughts expressed in words can be perceived by the outsider as completely incomprehensible and irrational.

I: But does another icon exist which can cover the rest of the mental illnesses—or, even better, all mental illnesses?

P: I don't think we will find anything with the impact of the madness icon. But we can find features common to all those states that we call mental illness. Otherwise, naturally, we would have no justification for using the term "mental illness." What would you propose as the distinctive features of mental as against somatic illness?

I: I sometimes think that the term "mental illness" must be a relic from the time when people assumed that a special mental substance was connected with the human body and susceptible, like the body, to its illnesses. Nowadays, virtually no one at all sees things in the old way. It is more a question of the mental illnesses being illnesses which afflict a human being's higher functions, especially thinking and the emotions.

P: What happens to thinking and the emotions when such an illness is present?

I: They are disturbed—or, if the worst comes to the worst, eliminated. The person cannot think lucidly: thought and speech become confused as in schizophrenia. Or the emotional life becomes exaggerated, inadequate, or utterly shallow as several affective conditions.

P: You seem to have worked in psychiatry. But let us take a closer look at how you express yourself. You say, for instance, that thinking and the emotions are "disturbed." You speak of the emotional life as being "exaggerated" or "inadequate." Surely, then, you must be comparing the sick person's condition with some norm, in relation to which this condition is disturbed, exaggerated, or inadequate? The $ 64,000 question which I ask you is this: What norm?

I: Oh, are we back to the discussion about norms again? I see that I cannot just refer to the population in general. We have to set a reasonable limit. I am looking for a straw at which to grasp. Let me start with survival again. A person who cannot think lucidly, or is even completely confused, has difficulties in managing the simplest things like procuring food and keeping house. Therefore, the gravely schizophrenic person's condition threatens her or his survival in the same way aș does the condition of the person gravely ill with cancer. It is also common knowledge that a risk exists that a deeply depressed person will commit suicide. Perhaps we can go on to analyze other states of mental ill-health in the same way.

P: But what about the manic? Is such a person in a danger-zone from the point of view of survival? Isn't such a person's condition extraordinarily productive?

I: I am begining to think that *you* are the naive one. In the first place mania, in a well-known variant of such disease, is often linked with depression, so the manic can be a depressed person too. In the second place, mania is almost always a fragile and transitory condition, so the manic is repeatedly confronted with "ordinary" reality. Manics find difficult to coordinate what they have thought and done in the manic state with what they think and do as "normal" people. Then, in the third place, the manic state often involves a total lack of realism. Manics repeatedly do mad things which land them in difficult situations with which they then have to cope and of which they have to take the consequences. Sometimes the things they do are of immediate danger to themselves and to others.

P: So the glad manic is not a person to be envied. Such a person's condition might not be normally life-threatening, but according to your description the condition does threaten the person's long-term welfare. Do you think that we can find a formulation which covers the types of illness we have exemplified?

I: Let us try the following: a mentally ill person is in a state which threatens either the person's survival or her or his long-term welfare.

P: Now we are well on our way to a general theory of health. Perhaps you are getting an idea of how we eventually could link this analysis to the analysis of the somatic diseases.

Nine

ON MENTAL ILLNESS, RESPONSIBILITY, AND POLITICS

I: You know, yesterday's discussion went far too fast and its conclusions were too simple. We said that the mentally ill person is in a state which threatens her or his survival or welfare. But how does that fit in with the history of psychiatry? I mean, the mad were not locked up just because they were a threat to themselves or their own welfare. Nor do we declare people ill, and call them psychopaths, for instance, just because their condition predestines them to a less than harmonious existence. Do we not also use the illness label to protect ourselves and the social order? The mentally ill person is a deviant who constitutes a source of anxiety in society and whom society therefore does not tolerate.

P: I am glad that you express yourself so directly. It clarifies the situation. I am curious as to how far you yourself were speaking.

I: I was mostly protesting against the simplified analysis toward which I saw you were moving. You ignore both the historical and the contemporary background. In the past, the mad and other mentally ill people were regarded as ill in quite another sense than the somatically ill.

P: You have a point there. A tradition of thought concerning mental illness has pursued paths all its own. If you like, we can begin with classical antiquity. For Plato and for many schools of philosophy that followed him, the health of the soul was the same as living a moral life. To be immoral was to have a sick soul. The Stoics embraced an extreme variant of this view. They considered immoral to have strong emotions, and thus they regarded wrath, alarm, terror, even great joy, as indicating a sickness of the soul. Because of this, the names of emotions became the names of illnesses.

During the period which you perhaps have in mind, the Modern Era in Europe lasting up to the middle of the nineteenth century (which Foucault and other historians of ideas have so vividly described), this total identification of immorality with mental illness did not occur. A person could be immoral, also criminal, without for that reason having the stamp of mental illness. The mentally ill did not receive the same treatment as the somatically ill. They were segregated from ordinary society, and nobody seriously thought of cure and recovery until the advent of psychiatry in the latter part of the nineteenth century. At the same time we must remember that mental illness was identified with madness. The "lighter" psychiatry is a modern addition.

I: But do you just want to assign that view to history? Aren't also many present-day psychiatric diagnoses "social" diagnoses? If manics or psychopaths are unbearable to their families or others around them, we take them in hand for the very reason that we find them unbearable. It isn't a concern for the health of the manic or psychopath which causes us to take her or him into custody.

P: No, cases do occur where taking into custody is warranted because a person is dangerous or otherwise behaves in such a way that living with that person is impossible. But from this does not follow that we equate mental illness with dangerousness or that we regard locking the person up as an adequate form of treatment.

I: But who are "we"? "We" are quite different persons. Look how things were in the USSR. There they were constantly locking people up for criticizing the social system. And they made broad use of the schizophrenia diagnosis. People who failed to accommodate to the system or who criticized Marxism-Leninism fulfilled the criteria of schizophrenia Soviet style. So it was critics of the system who were stamped as ill: political deviation equalled illness. Have we any reason to suppose that things are greatly different elsewhere?

P: Do you regard this to be the Swedish way?

I: Well, naturally not to the same extent. We are subtler and perhaps conceal the politico-pathological violence better. In a way, we have great tolerance. But I believe that if the basis of our political system, parliamentary democracy, were under threat, the illness stamp would present itself as a way of keeping disagreeable people in check. Just think of the difference in the way politicians behave toward "ordinary" political opponents and toward such radical opponents as question parliamentary democracy. The attitude is that these last are not fully responsible.

P: I grant you that there has been, and still is, a use of the illness stamp which is of a political, or for that matter moral, nature: the objectionable person is a sick person. We may speculate as to why such a use should still be tempting. I think the ascription of moral deficiency to the objectionable person ought to be more powerful, thereby making the person responsible for her or his views or actions. A person seen as being mentally ill cannot be held responsible: the person can't help it.

I: But doesn′t the power lie in this denial of responsibility? In the first place it gets rid of argument: to argue with a sick person is "meaningless". In the second place, we can pride ourselves on showing a touch of humaneness. We don't call for the punishment of the person; we call for the "treatment" of a sick condition.

P: You are right in that. But allow me to bring the discussion back to our main problem, finding a general characterization of mental illness. We had previously got to the point of regarding threat to life and threat to welfare as key concepts which we could use to elucidate the nature of mental illness. You retorted—and this was a good point—that other, quite different uses of the term "mentally ill" exist. But how does this affect our analysis? Do you think that modern psychiatry has something to learn from the point you were making? Should modern textbooks of psychiatry say that mental illness can also be a question of a condition which threatens parliamentary democracy, or a condition which the majority of the population dislike?

I: What your conceptual analysis comes to is that we require an *explication*. We need to produce a concept of mental illness which is reasonable, a concept fulfilling "decent" conditions.

P: You still see things rather polemically. I agree that we should perform an explication, that we should find a concept which is logically coherent and which we can use for its main intended purposes. But this concept has to be rooted in our basic feeling concerning mental illness. Let me also offer a retort to what you say about the use of the term "mental illness" as pure invective.

We now condemn the ignorant and above all inhumane way of treating mentally ill people which persisted even far into our own century. The code of ethics which the world's psychiatrists agreed upon in Hawaii in 1977 is extremely clear about the limits of psychiatry's field of application. Genuine psychiatry is not the handmaiden of politics. Soviet psychiatry was strongly condemned by the rest of the world's psychiatrists. Now that the USSR has fallen, Russia has denounced the degeneration and abuse to which the psychiatric profession lent itself. We cannot say, therefore, that Soviet psychiatry has been one of the accepted forms of modern psychiatry.

Nevertheless, I would admit that in an unconscious way every culture tends to set its own limits regarding mental illness. This is inevitable, though it should not be like that. The form of cultural impregnation I reject is that which is based on ideas of normality.

I: You're definitely no fan of the concept of normality.

P: Not in this connection. But my argumentation against normality as a basis for definitions of health does not derive from a fixed idea. I hope that became reasonably apparent when we were talking about somatic diseases. What causes me to argue against use of the concept of normality in the health context is that its supporters have such difficulty in deciding what the norm shall be—and even greater difficulty in justifying a chosen norm.

Things are especially delicate when we come to mental illness. A psychiatric theorist in the USA has wondered in all seriousness whether the Americans, extremely influential as they are in psychiatric diagnostic theory, are not in the process of setting up American middle-class behavior as the norm of mental health. Serious deviations from the conventions which regulate this behavior risk being stamped as symptoms of mental illness. But why should these conventions be normative? And what happens to international scientific communication? If diagnostics is to be based on the American middle-class ideal, this has to be spelled out. If, however, it is to be completely culture-relative, we have extraordinarily little chance of conducting a meaningful international dialogue.

An extremely large and difficult research task is to elucidate how these things are in practice. The investigation of diagnostics I referred to in our second conversation is just an isolated example, dating from the beginning of the 1970s and not directly concerned with the question of normality. It showed in quite general terms that diagnostic criteria can be different from one professional culture to another. However, I believe that psychiatry can be, and indeed should be, culture-invariant with regard to questions of normality. I think it an error—and here I am sure that I have the international psychiatric establishment on my side—to call upon psychiatry to treat people of deviant behavior irespective of the type of deviance. That would mean that all criminal behavior, all immoral behavior, all unconventional behavior, at least in their most serious forms, would be the subject of psychiatric treatment. But nowhere in the world is this the state of affairs. We still have our prisons, we still censure our fellow-beings for their immoral behavior without declaring them to be ill, and we still allow a few eccentrics to exist in our midst without summoning the aid of psychiatry.

I: But some forms of wholly "deviant" behavior do have diagnostic designations. I am thinking chiefly of sexual deviation. Most Western societies have until quite recently classified homosexuality as an illness, and still classifies other forms of sexual deviancy as illnesses.

P: You have taken up something interesting which will make our further reasoning easier. The position of homosexuality as a form of illness has in recent decades given rise to a debate, first and foremost in the USA, which has resulted in significant clarifications within psychiatry as to what shall constitute general criteria of illness. But let me briefly tell the story.

In the 1960s, the homosexual political movement in the USA, the Gay Liberation Movement, became extremely strong and increasingly radical. The members of this movement were no longer satisfied with the humanitarian gain which had been made when homosexuality was declared an illness and was no longer criminalized. They had strong doubts concerning the reasonableness of the illness label, and they vigorously asserted that this label contributed to the disparagement of, and discrimination against, homosexuals on the part both of individuals and of society. The pressure of society was such that the majority of homosexuals had to hide their homosexuality. This included homosexual psychiatrists. Had they declared their homosexuality, their careers would have been ruined.

During the period 1970-1973, the Gay Liberation Movement acted with extreme vehemence and aggression. Most effective was their conduct at the large annual congresses of the American Psychiatric Association. They disrupted lectures on the chemical and behavior-therapeutical treatment of homosexuality; they demanded to be present at the discussions on this subject; and they declared that many of the psychiatrists had a distorted perception of such things inasmuch as they had only come into contact with homosexuals who had sought their assistance in the solution of problems. The psychiatrists know nothing, asserted the Gay Liberation Movement, about the life of well-adjusted homosexuals.

The movement's activities had an unexpectedly strong impact. The American Psychiatric Association decided to conduct a thorough investigation of the question of the status of homosexuality as an illness. They called in a group of experts to work out a basis for an official resolution. One of these experts, the psychiatrist R. L. Spitzer, came to play a decisive role with regard to the group's report. The fascinating thing was that Spitzer took the matter of definition seriously. He expressed the opinion that the question of the position of homosexuality as an illness is not a political one. Nor can we resolve it within the perspective of any of the different psychological theories, for instance, that of psychoanalysis. We must be in agreement about the psychopathological criteria, about the general grounds upon which we decide whether to designate a condition as pathological. Spitzer suggested—and in this he had the committee with him—the following. We are to regard a condition as a mental illness—or "mental disorder," to use the broad term which came into general use—only if the person suffers and/or the condition involves a substantial restriction of the person's social ability. If neither of these criteria is fulfilled, the condition is not a mental disorder.

So what conclusion did they come to in the case of homosexuality? That homosexuality did not in general fulfill either of the criteria. Certainly many homosexuals do suffer because of their homosexuality, but this is principally because of the reaction to it on the part of those around them, not because of the homosexuality itself. Certainly, too, many homosexuals are socially handicapped, but here again this is principally because of society's attitude. In the opinion of the group of experts, no grounds existed for maintaining that homosexuality in itself involves (as does, for instance, schizophrenia in itself) subjective suffering or social handicap. So no grounds existed, either, for classifying homosexuality as a mental disorder. The group's opinion gained assent at the 1973 congress of the American Psychiatric Association, and homosexuality disappeared from the American list of illnesses. A disease was wiped out by definition! In Sweden, the National Board of Health and Welfare passed a corresponding resolution in 1979.

Spitzer was not only a member of the group of experts but also one of the main editors responsible for the revision of the American *Diagnostic and Statistical Manual of Mental Disorders* (*DSM*) which was carried out in the 1970s. The new version was *DSM III*, but since then *DSM IV* has appeared. In the introduction to *DSM III*, we find the main criterion of mental disorder which Spitzer had proposed in the course of the debate on homosexuality. What conclusion can we draw from this? Well, we can say that serious analysis has been carried out within psychiatry itself and has led to a clarification of the criteria of illness. These criteria have to do with the subject's suffering and disablement. They have nothing whatever to do with any deviation from normal behavior. We might say that I am in good company when I reject the idea of a social norm as the basis for defining mental illness.I: Thus, we are back to suffering and disablement. I suppose for the moment that I must admit defeat. Let us see, then, if your program for health analysis can take us any further.

Ten

ON THE LATITUDES OF HEALTH AND ON THE DIVIDING-LINE BETWEEN HEALTHY AND ILL

I: Up to now you have mostly done no more than hint at the paths to a characterization of health. Now I must ask you to be more systematic and to keep yourself on a tighter rein.

P: Gladly. But I don't want to be needlessly formal. I think our best starting-point will be the general icon of disease which we discussed in our previous conversation. I believe you became fairly convinced of its reasonableness. From now on I will call it the ill-health icon. We must not allow disease to represent the entire antithesis of health.

I will also be looking upon health as a dimension, extending from a state of ideal or perfect health to a state of maximum ill-health. I regard as misleading to speak of health and ill-health as two mutually exclusive and discrete states. Obviously that doesn't correspond to reality, where health varies enormously in degree.

I: On that point I agree with you completely. But one thing I am curious about: what *is* maximum ill-health? Is it the same as death?

P: My answer to that question will emerge when I attempt a more precise characterization of health. We will find that death is not the only state which constitutes maximum ill-health. But let me now represent the health scale graphically by means of a line whose two ends represent optimum health and maximum ill-health.

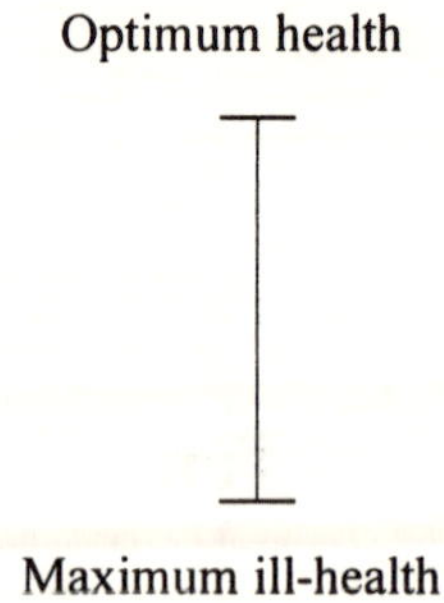

I: Is one line enough? Aren't there variants of ill-health? Ought it not to be possible to distinguish between people with the same degree of ill-health but enormously different types of ill-health? Compare, for instance, a polio victim and a schizophrenic. They can be "equally" ill in the sense of the seriousness of their afflictions, but the two types of ill-health are immensely different.

P: You are quite right. We need to make many distinctions. The huge difference between the polio victim and the schizophrenic is not only a question of type of illness, of the difference in the grounds of their ill-health, but also a question of which functions are affected. The two people are handicapped in totally different ways. But let us postpone our consideration of the diseases and keep to the health scale for a while. I start by representing the health dimension as *one* scale to indicate the comparison of degree that we are inclined to make between different states of health. We say that A is healthier than B. We say that B is more ill today than yesterday. So we conceive of a general grading of states of health which we can perform without going into detail. From the theoretical point of view, starting from this dimension is necessary. We can go into greater detail on the basis of it. Let us have a look, then, at that line going from optimum health to maximum ill-health. Can you tell me what you think the figure chiefly lacks?

I: What about the transition? When does health pass into ill-health? How are we to mark the position on this continuous scale?

P: Well, there we have a central question, one which many thinkers regard as the most central question. Where does the dividing-line go between healthy and ill? When have we the right to declare a person to be healthy, or ill? And if we look upon health as a continuous dimension, where on earth can we reasonably draw the magic line?

I want to defuse this issue by means of a backward glance. I enlist the aid of Galen, the great physician and philosopher of classical antiquity who wrote a large amount about anatomy and other basic medical subjects but also about the theory of health. Galen had a whole theory about what he called the latitudes of health. He assumed, just as we have done, a health dimension extending from optimum health to maximum ill-health. But he also made other assumptions with which we feel less at home. Galen did not just look for a dividing-line, somewhere in the middle, as we are inclined to do. You see, he did not presuppose, as we do, that people are either healthy or ill. He was of the opinion that people can also be in a *neutral* state, neither of health nor of ill-health. In the form of a figure, Galen presented his theory like this:

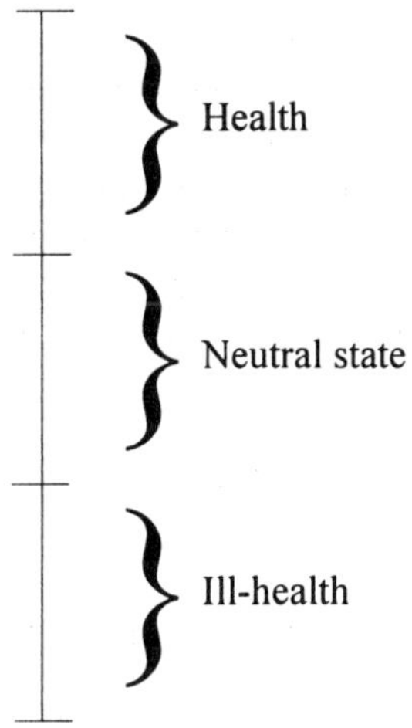

I: But what on earth is a neutral state? When do I feel neutral?

P: You show by your questions that you are quite fixed in your view of the health dimension as having two parts. To you, health and ill-health are contradictory concepts, while to Galen they were contrary ones, if we are now to use the terminology of logic.

I: The difference between contradictory and contrary you'll have to explain.

P: I can explain it with the aid of simpler concepts. Black and not-black are contradictory concepts, while black and white are contrary ones. Two concepts are contradictory if together they exhaust a complete dimension, as in this case the color dimension. This is not the case, however, with the contrary concepts. Obviously an object must be either black or not-black. But it need not be either black or white, for it can be yellow or green or whatever. So when we say that we *must* be either healthy or ill, we are treating health and ill-health as contradictory concepts. Galen saw things differently.

I: I suppose Galen would have been able to provide a concrete description of a neutral state.

P: Yes, I am sure he would have been. For Galen, the person who is neither manifestly ill and disabled nor bursting with energy, is a person in a neutral state. I can imagine that your own state on a typical Monday morning fits quite nicely into Galen's neutral category.

I: Might it be the case, then, that most of us are in a neutral state most of the time?

P: It is possible. Galen's schema, you know, was in use for considerably more than a thousand years in Western medicine. Especially in late-medieval Italy, it was the subject of many scholastic expositions. The application of it must have varied a great deal from one period to another and from one physician to another. For some, the neutral category must have been extremely large, just as your question suggests. For others, it represented a buffer zone between healthy and ill.

I: I find fascinating that this idea of a neutral state has been applied in practice.

P: Yes, and for such an enormously long time. Galenic medicine takes up a much larger part of recorded history than does post-Galenic medicine.

I: Did Galen make any other additions to the health dimension?

P: Yes, he added a further refinement. He considered that a person's state of health has two levels, at every moment. One is the basic, the level we are concerned with when we say that someone is basically healthy. The other is the temporary: you can be healthy at the moment, or ill at the moment. The two levels can then go together, for example, in the following way. We can say that a young girl is basically healthy but at present has an infection of the upper respiratory passage. Conversely, we might say that an elderly person with cancer is basically ill but nonetheless at present in reasonably good health.

I: Are we to include the neutral states in this schema too?

P: Yes, we are. You can be basically neutral but at the moment ill or healthy, and so on. So nine combinations are possible:

Basically healthy	Healthy at the moment Neutral at the moment Unhealthy at the moment
Basically neutral	Healthy at the moment Neutral at the moment Unhealthy at the moment
Basically unhealthy	Healthy at the moment Neutral at the moment Unhealthy at the moment

I: The idea of "basic health" and "temporary health" is a perfectly reasonable one for which we often have a use, but I have never come across it systematized like that before.

P: No, remarkably enough we aren't more familiar with the history of ideas. But this was just a little digression. Let's get back to our main discussion, about the dividing-line between healthy and ill. My assertion is that the line is not a theoretical one in the sense that we can discover or otherwise establish it with the aid of observation or pure deduction. Principally the line is a question of the minimum degree of health which the person finds acceptable. The moment I find my present state of health unacceptable, I have crossed the line into ill-health. That moment is also typically, though by no means always, the moment when I begin to think of seeking medical help.

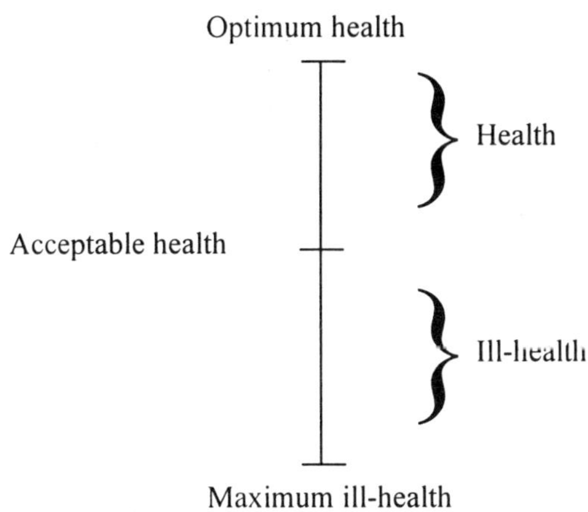

This is the basic function of the concept of acceptable health: it enables us to draw a line between healthy and ill. Then we can imagine other dividing-lines based on this. One such line, which we can call a social dividing-line, separates the acceptability and the non-acceptability of the person's condition from the point of view of health-care staff or others with whom the person comes into contact. My position on this is that the person's own assessment of her or his health should be the starting-point for the assessment which health-care staff and others make.

I: But do you think in all seriousness that the individual should define what is an acceptable state of health and what is not? I mean, the dividing-line can vary enormously from one person to the next. Some people are patient and wait for ages before they report sick, others, their opposites, won't put up with the least little ailment. And what is to happen in the case of those—children and the retarded—who lack the ability to assess their own health?

P: First a necessary distinction. When I speak of a person's assessment I am not equating this with what the person *says*. A discrepancy can exist between the two, as I use the term "assessment". The most extreme discrepancy occurs when someone tells a lie. We can lie about our state of health just as we can lie about anything else. An untruthful statement cannot indicate the dividing-line between acceptability and non-acceptability. As physicians are especially aware, people can exaggerate the degree of their ill-health in order, for example, to gain advantages.

So the assessment I am speaking about is of a more abstract sort. It is a matter of what state of health the person would find acceptable if she or he were perfectly truthful and also put a lot of thought into making the assessment. You are right, though, insofar as large individual variations can occur even when these two conditions have been met. People have such different expectations concerning life. Some expect to have the more ideal sort of life, others have much lower demands. But—and this is my point—health is to some extent dependent upon the person's level of expectation. The more you ask of life, the greater the chances of your falling into a state of ill-health. This is not just an empirical assertion but also a conceptual one. Health, according to the analysis which I am advocating, is an individual-related concept.

I: I feel completely confused. Do you mean that a person's ill-health, or in other words the presence of this or that disease in the person, is dependent upon what the person asks of life?

P: Nice that we can finally get to grips with that question! You have to distinguish between ill-health and the presence of disease. I have attempted to make this plain before, but the distinction has remained in the wings during today's conversation. I have also contributed to keeping it at a distance with my talk of a dividing-line between healthy and ill. The dividing-line I have intended is that between health and ill-health. We have not so far seriously introduced the diseases into the systematic discussion. What the person assesses is the degree of health. When the question concerns whether a disease is present, often she or he can say nothing.

I: It is rather unusual to differentiate between disease and ill-health in that way. Is it truly necessary? And how does it fit in with your conceptual-analytical program? Wasn't ordinary language your starting-point? And surely you have to agree that in ordinary life we often think of disease and ill-health as the same thing?

P: I appreciate that you press me on this point. People often find it puzzling. So let me explain it as carefully as I can. Naturally, I agree with you that in ordinary speech we often make no distinction between "Jack has a disease" and "Jack is in poor health." But if we think about it, we see that a distinction does exist. I mentioned in one of our earliest conversations that ill-health can also be due to injury or defect. People working in health care, especially primary care, now also recognize that a person can be in a state of ill-health even when no demonstrable disease is present. From this we can conclude that our ordinary feeling for the language is an uncertain guide on this point. In a conceptual analysis, therefore, we must perform the explication which is most fruitful for theory. I find considerably more fruitful to use the term "ill-health" in the more comprehensive sense, whereby ill-health can be due to any number of factors, the most typical of them being disease.

I: So you think that ill-health can be due to other factors than the three you mentioned (disease, injury, and defect)?

P: Yes, I certainly do. Ill-health need not be linked with disease, injury, or defect. For instance, it can be a question of existential problems. However, the fact remains that disease, injury, and defect are the most typical causes of ill-health. I also regard them as conceptually related to ill-health, though not as strongly as you were inclined to assume.

I: But you have sidestepped one of my earlier critical points concerning your individual-related concept of health. You related the dividing-line between acceptable and unacceptable states of health to the individual's acceptance or lack of acceptance of her or his own state. But what if the person is a small child or suffers from senile dementia, or for that matter is a person we outsiders consider psychotic? Are we to trust the word of such people when we are attempting to assess their respective levels of acceptable health?

P: You're sharp-sighted and don't let anything pass by unnoticed! But let me remind you of some of the qualifications I have just made. I have not equated the assessment of health with what people say. I spoke of the state found acceptable on the basis of proper reflection and truthfulness. That's not applicable to people incapable of reflection, incapable of assessing their own situation. Small children, people suffering from senile dementia, and psychotics are examples of people who are more than likely to lack the ability for the holistic reflection required. In the case of such people, an outsider will have to make the assessment. Note, though, that the outsider must make this assessment from the perspective of the person concerned. The outsider must make every effort to identify with the person, to experience the person's state of health from within. Why this should be necessary will become more apparent later, when I elaborate my view of the concept of health.

I: I realize that I need to know more about your system before I can take up a position on this. Therefore, I would be glad if our next conversation you give your view of the nature of good health.

P: Yes, you must look upon this conversation as just a preliminary one. Some of the terms I have used here, like "expectations concerning life," I will replace with more precise and well-defined ones. If you have serious criticism to offer, it should concern what is to come and not the preliminary reflections I have set forth in this conversation.

Eleven

HEALTH AS ABILITY TO ACT

P: In previous conversations we have in several ways laid the foundations of a systematic characterization of health. Some of these ways have been negative in that they have involved rejecting earlier attempts at such a characterization. We rejected the attempt to define health in biostatistical or bionormative terms. We then attempted to find the typical everyday icons for different sorts of ill-health. We found rather different icons for somatic and mental ill-health, but afterwards agreed that they most likely had a lowest common denominator. In the conversation before this one, we established the position of health as a dimension and in doing so distinguished between the concepts of optimum health and minimum health. We also drew a distinction between the health/ill-health scale and the concepts disease, injury, and defect.

I: Shall we now take the ill-health icon as our point of departure? Is ill-health the same as suffering and disablement? Is health no more than the absence of these?

P: Roughly, yes. But an adequate characterization requires more. I am sure that you can see difficulties attaching to the definition you propose.

I: Yes, indeed I can. To start with, not all suffering is ill-health. For instance, I cannot accept that grief is ill-health. Nor, generally, is disappointment or rage ill health, though both these emotions involve suffering. So we have to find the right sort of suffering.

P: Suffering can come in many forms, and not all of them have directly to do with ill-health. Most of your examples of suffering have to do with emotions. Emotions are feelings which contain an element of reflection. To put this in technical language, emotions are cognitive concepts. Take, for example, the emotion of disappointment. When you are disappointed, you are disappointed about something of which you are thinking or have thought. You can be disappointed at having failed an examination. This disappointment presupposes, naturally, that you are aware of the examination failure. The simpler feelings, which we call sensations, are quite different: they presuppose no reflection. Typical examples are pain and nausea: to feel pain or to feel sick, it is not necessary to think of something or know of something.

I: Then suffering of the sensation type is what distinguishes ill-health?

P: That is typical, especially with somatic ill-health. Things are more complicated when we deal with mental ill health, but we need not go into that now. It isn't necessary for my purposes in this conversation. You see, I don't intend to use suffering in my basic definition of health and ill-health.

I: How can you say that? I mean, pain and other forms of suffering are the most typical phenomena of ill-health. Unless I misunderstood you, you have accepted suffering as part of the ill-health icon.

P: Yes, I have. Suffering forms a part of our most common notion of disease and ill-health. But let me repeat: when we are attempting a characterization which is systematic and as universal as possible, we have to think carefully. You yourself made the point that not all suffering has to do with ill-health. I now wish to make the point that sometimes ill-health is not directly connected with suffering, at least not suffering on the part of the person concerned.

I: What extremes have you unearthed now?

P: A perfectly plain and not especially extreme case is unconsciousness. Other cases are ones to do with certain injuries and defects. Some forms of paralysis cause no pain whatsoever; people with some types of mental illness do not suffer.

I: But what have these cases in common with other forms of ill-health?

P: We must concentrate, I think, on the other element of the ill-health icon. I believe that we can find the essence of ill-health in disablement or inability to act. This is the concept which offers us the best chance of achieving a universal characterization of the field of ill-health. Obviously, unconsciousness and paralysis involve an extremely high degree of inability. But a strong link exists between inability to act and suffering. Suffering leads to inability to act. If you are in great pain, then you cannot undertake anything at all. If you feel sick, your ability to act is also severely limited. In brief, inability to act appears to cover the whole of the relevant field, whereas pain and other sensations of suffering have a more limited range. This is not to deny the enormous role which pain and other suffering play in the lives of many people in a state of ill-health.

I: So inability to act, or disablement, is the key concept in your characterization of ill-health.

P: Exactly. The distinctive feature of ill-health is an inability, either directly attributable to a somatic phenomenon, such as paralysis, or indirectly attributable to pain, nausea, or other suffering.

I: Do you mean, then, that mental ill-health also in general has to do with inability?

P: That is my hypothesis, and Spitzer and DSM III appear to agree with me. No shred of doubt exists about it with regard to depression, schizophrenia, and most neuroses. But also mania, which you might think was a counter-instance, involves a large social inability. It is characterized by a lack of control over our own person, which can have the most disastrous consequences.

I: But then comes the big question: When does inability exist, and how and by whom is its existence established? Is it a social assessment or do more stable criteria exist?

P: You have brought up an extremely interesting question about which I have spent a lot of time thinking. It is perfectly correct that we can never content ourselves with speaking about inability or disablement. Here DSM, the manual which I mentioned before, has a long way to go. We still have to set inability in relation to a norm, to a guideline representing a level of good health. For us to regard the person as having reached this level, she or he must in the most explicit way be able to see, hear, speak, understand, and perform many other actions. If we have no such minimum level of good health, then we just cannot speak about inability in any of these respects.

I: You are back to the norm question. Do we have to go through the normality arguments all over again?

P: No, I hope that our earlier discussion will have ruled out the idea that statistical normality should constitute the norm for human ability. Nevertheless, we do logically need some norm or guideline. If we say that a person lacks ability in a particular respect, we must be able to indicate what the corresponding ability should involve.

I: But how do we decide on the guideline? Is it up to society? And how do we decide whether the guideline is reasonable?

P: No, it is not up to society. As you are well aware, no authorities establish in detail how much a person should be able to see, hear, speak, understand or otherwise perform in order to be considered healthy. (Forming an exception are the requirements in the case of those who are to perform special types of function: pilots, naval officers, firefighters, etc.) We find no such indications even in the 1980 WHO classification of handicaps, which is the document where we might most expect them.

I: But cannot we say that society, in the form of the physician who puts a person on the sick list, does make a decision as to what this or that level should be?

P: Though the physician may make that decision in the individual case, she or he sets up no norm for the entire population. Furthermore, the physician's decision is not arbitrary. In the typical case, the physician talks things over with the patient, and the two of them together decide what care or treatment the patient needs.

I: So is ill-health a state where the person has need of care or treatment?

P: That is a good starting-point for thinking about it. Mr. Smith is in a state of ill-health when he, for reasons deriving from his own person, cannot cope unaided. Note that the question is of the person's own situation. Mr. Smith's situation may be vastly different from the situation of Mrs. Jones. You can easily picture how different are the demands of everyday life in the case of, say, a head of state, a bishop, a farmer, a metalworker, a sailor, a cook, and an occupational therapist. Precisely because of this variation, it is impossible to set up any completely general norms with regard to what the healthy person should be able to do in different respects. What a healthy head of state must be able to do differs from what a healthy metalworker must be able to do. In some respects more is demanded of a head of state, in other respects more is demanded of a metalworker. In each case she or he faces distinctive challenges.

I: Do you seriously mean, then, that the norms of health are completely individual?

P: But that's hardly revolutionary? It must be the model that the wise country doctor and district nurse have followed throughout the ages, especially with regard to rehabilitation. Rehabilitation is a question of enabling a person to return to a level where she or he will be able as far as possible to go on living as before.

I: But that means that, strictly speaking, it is impossible to establish whether a person is in a state of good health or of ill-health without knowing an enormous amount about the person. We must have an extremely good knowledge of the person's situation and the demands it involves.

P: Splendid. I shall soon be able to offer you a job as my information officer! It is at any rate impossible to establish the presence of good health without first becoming acquainted with the person's situation and its demands. However, we can, especially in grave cases, establish that ill-health is present without the aid of such all-round knowledge.

I: How do you mean?

P: Consider, for instance, the person who is unconscious or who for some other reason cannot communicate with those around, or who is completely unable to move. In such cases, we can say with certainty that the person is in a state of ill-health for the simple reason that all conceivable situations in life require such an ability as the person lacks.

I: So we might speak of a universal human minimum. People who fall below this level must be in a state of ill-health no matter what their situation and its demands.

P: Yes, such a minimum exists, even though it is difficult to precisely demarcate it. A basic ability of communication and movement must be a minimum requirement. Then comes an extensive zone within which most people need a greater ability but where a few people do not—people who live extremely sheltered and undemanding lives and who are able to manage well enough with rather limited ability.

I: Something is unsatisfactory in the statement, "My situation demands ability X," if we are to use it as a criterion of whether the person is in a state of health or of ill-health. It has something static about it. Are we always necessarily to set people in relation to the way their lives are at the moment? What if the person's present situation is wretched and needs changing? What if the person is an alcoholic or prostitute? It would be absurd to allow the abilities required for living such a life be the standard for judging the state of the person's health.

P: Certainly. People's situations can be miserable. People can also, even when their situations are not miserable, have the ambition to change their lives, to develop themselves. This also has to be taken into account, within reasonable limits, in health care. Therefore, in my formal characterization of health I do not speak of what "the situation demands," although as a step in our reasoning it has functioned well enough.

I: Then what expression do you prefer?

P: I suggest the expression "vital goal." A person's vital goals include much more than the situation demands (to come back to that expression). The reference to a person's vital goals enables us to take account of her or his ambitions concerning change. At the same time, the vital goals include the ordinary demands attaching to the present situation. Generally, people have the following vital goals: to preserve family ties, to go on living in their present home, to keep their job—but also such things as to complete their education and improve their job situation. The vital goals include both static and dynamic goals.

I: But what is a vital goal?

P: The vital goals are the goals which are of especial importance to the person to attain or, if already attained, not to let slip. If the person cannot attain these goals, her or his satisfaction with life will be affected.

I: So must the person attain the vital goals in order to be happy?

P: Yes, we might say that, although we are speaking of a minimum happiness. There is an immense amount which we should like to have, and which, if we should get it, would increase our happiness. We can wish for, and perhaps obtain, a pay rise, a letter from someone dear to us, a bit of sunshine for our holiday, and so on. But not all such things constitute vital goals. Not all such things are prerequisites for our continuing to live a tolerable life, for ensuring our minimum happiness.

I: Are you after a basic need?

P: To speak of a basic need is a step in the right direction. The problem is that, as several researchers have shown, the concept of need does not provide us with a full answer. To say that someone has a need is to say something incomplete. If we have a need, we have a need *of* something (the object of the need) *for* something (the goal of the need). For instance, we have a need of food for survival. So survival is the goal of this need. Perhaps in speaking of a basic need you are thinking of the need for survival. But this must be insufficient if our purpose is to characterize health. As we have discovered through our discussion, health must be something more than the ability to survive. I wish to capture this "something more" by means of the vital goals.

To be healthy, we must have the ability not only to survive but also to live a tolerable life, given reasonable circumstances.

I: What are these circumstances that you have introduced into the definition of health?

P: Nothing is odd about introducing them. Every time we describe an ability or inability we must say under what circumstances it applies. For instance, you can't go on reading under any circumstances whatsoever. If you are constantly disturbed by noise or by someone claiming your attention, you give up. Nor can you go out in a rowboat under any circumstances whatsoever. If a violent gale is blowing toward you, you will get nowhere. Yet you still say that you have the ability to read and to row. When you say this, therefore, you must implicitly have some circumstances in mind. We commonly speak of "normal" circumstances, but I think we should avoid that in our discussion.

I: So you chose to speak of "reasonable" circumstances instead. What do you wish to indicate by that?

P: I wish to indicate that if we attribute health to a person, then we are maintaining that this person can attain her or his vital goals within the framework of circumstances which we regard as reasonable. This does not have to mean that the circumstances are statistically common; we encounter statistically uncommon circumstances which are not unreasonable.

I: But "reasonable circumstances" is not an enlightening term.

P: It is enlightening in that it says that the demarcation of the class of circumstances is not given from the start but is established normatively by the speaker. So here I have a significant reason for speaking of health as a value-related concept.

I: But if the speaker determines the circumstances, how should physicians, or people in general, communicate with one another? Must we not together decide that we should regard a class of circumstances as being reasonable?

P: Perfectly correct. Let me clarify what I am attempting with my characterization of the concept of health. My starting-point is a concept which is given in ordinary language. It is a concept which from the beginning has never received an exact interpretation. In several respects, it has open borders. In my basic analysis, I wish to retain much of this openness, though at the same time I hope that I am accomplishing clarification, for instance, in indicating the features of the openness.

The concept of health has everyday uses where the problems you indicate arise. Speaker S says that Mr. A is not completely healthy because he is unable to do his job properly. Speaker S_1, however, regards A as completely healthy and explains A's apparent inability with reference to circumstances at A's place of work which S_1 does not find reasonable. S and S_1 do not agree in their judgment of what are reasonable circumstances. But such problems can be avoided, and should be avoided, in health care and in health-care research. We can sharpen the existing concepts in different ways. To this subject I shall return.

I: I hope you will. But now I am curious to hear your precise definition of health.

P: Well, I propose the following definition: A person P has optimum health (is completely healthy) if, and only if, P is so constituted (biologically and psychologically) as to have the second-order ability to realize, given reasonable circumstances, all her or his vital goals.

I: I see that your definition brings together several of the basic concepts which we have been discussing. But I wonder about the weird newcomer. What sort of creature is a "second-order ability"? Why not "ability" pure and simple?

P: For all ordinary purposes, "ability" is sufficient. But in some situations we run into difficulty if we only take into account a person's present ability. Let me show you what I mean. A ten-year-old boy has for some reason never learnt to ride a bicycle. The family moves to a remote area where it immediately becomes vital for the boy to be able to cycle to school. During a brief transition period, before he has learnt to ride, the boy is handicapped, and one of his parents has to drive him to school. It would be odd, though, to describe him as not being completely healthy. He does have the basic ability to learn to ride a bicycle; in other words, he has a second-order ability to do so.

We can find many similar examples in the life of an immigrant. A person coming to a new culture suddenly lacks many relevant abilities. She or he cannot speak the language, is not familar with many of the new country's conventions, is not familar with legal conditions, and so on. So the immigrant may be severely handicapped in the new situation. Nonetheless, the person most often has the ability to learn such things. Thereby she or he has the second-order ability to cope with life in the new society. So we can ascribe health to the person.

I: Yes, it would be absurd to say that the immigrant was not in good health just because she or he was unable, for cultural reasons, to cope in the new environment.

P: You agree with my line of thought, then.

I: Let us say that I am trying to follow it. At the moment, I would be going too far to say that I agree. I must submit your ideas to a much more severe examination.

P: Please begin doing so in our next conversation.

Twelve

MORE ABOUT VITAL GOALS

I: It takes time to digest all your analyses and proposals. I can see now why you have settled for vital goals and have rejected the idea of basic needs as well as the idea of what the situation demands. But we need to take a much closer look at the notion of vital goals. A first essential question is whether such goals need to be conscious ones. When you speak of vital goals, do you mean ones that people consciously set up for themselves?

P: Many vital goals will be conscious, and the majority are intentional. But we have to differentiate between what is conscious and what is intentional. Intentions can be unconscious. Freud was able to show that we are not always conscious of what we intentionally do.

I: But a lot of people never set up any goals in life, or just set up trivial and short-term ones. They perhaps think only of their needs in the immediate future, such as getting food, stopping the rain from coming in, or defending themselves against an intruder.

P: Yes, many people do not have conscious goals other than ones of that sort. But we can easily show that they also must have basic intentions. They virtually always have the basic intention to live as long as possible and to live as tolerable a life as possible. Usually they also have several other such intentions—for instancc, to go on living in their native country, probably in the part of it where they were born and grew up, and to go on living with their nearest and dearest.

I: But we can find people who don't have even these basic intentions? I am thinking of small children and the retarded. Often they cannot imagine things beyond what is immediately in front of them. It is far-fetched to ascribe to them the intention to live as long as possible or to go on living in the place where they were born.

P: Yes, they can scarcely be said to have long-term intentions. Nevertheless we would be wrong to refuse to ascribe to them any intentions at all. Intentional behavior is apparent even in the case of a tiny infant. The infant grasps at something it wants; it defends itself against something that frightens it; it seeks the warmth and tenderness that its mother can provide.

I: So would you say that the infant, too, has vital goals?

P: Yes, I have to say that, otherwise I can have no hope of my theory being general. The infant must be able to grasp, must be able to suck, must be able to turn around, if we are to call it healthy.

I: But what is it, more precisely, that distinguishes the vital goals? We have many intentions, but some of them are all too thin and trivial. They can't all fulfill the criteria of vital goals, if your theory is to be interesting. Another thing: must all vital goals be matched by corresponding intentions?

P: Not all our intentions are directed toward vital goals. We have, as you say, many trivial intentions. Nevertheless, we need to draw a distinction between different sorts of triviality. When we speak of trivial intentions and actions we sometimes mean the intentions and actions of everyday life. But many of these are vital. It is vital for us to eat, to be warm in winter, to rest at night. These everyday goals are logical consequences of our vital goal of survival. They are means of survival, and therefore we can call them derived vital goals. But not all everyday actions have such a central role. We do small things all day long that are not parts of a more long-term plan. All such things lie outside the vital goals, basic or derived.

You asked whether all vital goals have to be matched by corresponding intentions. The answer is no. Some people can have vital goals which they do not even unconsciously intend to attain.

I: Now I am confused. Considering what you have just said, I wonder why we bother with the term "goal" at all. And how can we ascribe a goal to a person which the person does not even intend to attain?

P: I think you are caught up in the (admittedly common) use of the word "goal" to refer to an end which we consciously and intentionally seek. My concept of a goal, however, does not imply consciousness or intention. At the same time, I have pointed out that in the typical case a person's vital goals are also intentional ones.

I: What, then, is your formal definition of the concept of vital goal?

P: It is this: a vital goal for the person P is a state of affairs in the world which is such as to constitute a prerequisite for P's long-term minimum happiness.

I: Well, that is something for us to get our teeth into. Let us begin with "state of affairs in the world." Do you mean anywhere at all in the world? Can I have a vital goal which has to do with a state of affairs in Greenland or Galápagos?

P: I do not want to exclude that. That you should have a vital goal which had to do with Greenland or Galápagos is highly unlikely; nonetheless, I can describe a case where that could occur. Let us imagine that you are a development assistance worker in Galápagos, that you make a lot of friends there, and that you become deeply involved in how things go for the islands. You believe you can influence the situation there, and you strongly wish to do so. In such a case for you to form a vital goal which has to do with Galápagos is completely natural.

I: You speak of "forming" a vital goal. It sounds like a conscious decision.

P: It *can* be a conscious decision but need not be. For instance, a vital goal can come into being through a person's falling in love, and falling in love is seldom the result of a decision. I do agree that it is possible to interpret the word "form" as implying an intentional act. Sometimes language is inadequate when we are trying to express something exact in a concise fashion.

I: When you speak of falling in love, I see that it must be possible for a vital goal to be temporary. I mean, people fall out of love too.

P: Yes, the concept of a vital goal involves no presupposition of durability. The most basic vital goals endure for the vast majority of us, while other vital goals are temporary. What makes things complicated is that we can describe the vital goals on different levels of abstraction. Love (in the abstract) might be an enduring vital goal for us. (I am inclined to think that it is a universal enduring vital goal.) Being in love with this or that person, however, may be a more temporary vital goal.

I: Now the word "universal" has entered into the discussion. What is the situation with different people's sets of vital goals? Are the sets much the same, or quite different, or what?

P: That is a question which in the first place we have to investigate empirically. Nonetheless, we can say something about it on the basis of experience and with the aid of a little thought. The basic vital goals, those which concern sheer survival, are universal. Probably some of the other vital goals are universal too, or virtually so. But we can also find large individual differences. This is above all the case, as I indicated just now, when we come to describing the goals more concretely. We can easiest see such differences if we consider the enormous range of occupations people have. For the captain of a ship, a vital goal is that the ship shall safely reach port. For a builder, a vital goal is that the building go up. For an author, a vital goal is that the book be written.

I: The dissimilarity between people's sets of vital goals has consequences for the concept of health. Good health for the captain of the ship is not the same as good health for the builder or the author. Isn't that a strange consequence? Shouldn't health be the same for all of us?

P: What makes you think it a strange consequence? Think of the purpose of medical rehabilitation. This purpose is to restore people to a position where they can can take care of themselves, their household, and their work. (Other purposes occur too, but allow me to simplify.) So rehabilitation has to take into account what occupation the person has. If a skill is indispensable in the occupation in question, it must be specially trained. So the goals of rehabilitation must be different from case to case. Complete health must imply different things in the different cases.

I suspect that your wonderment is attributable to your seeing in front of you a misleading picture of how this rehabilitation is to take place. You may have before you an image of special rehabilitation centers for captains and authors respectively, where the whole focus is on the distinctive characteristics of the one or the other occupation. But that is nothing but a caricature. We must remember that the captain and the author have a great deal in common; they are probably in agreement when it comes to the majority of vital goals. A captain, too, needs to be able to write. Furthermore, the captain may happen to have a great interest in literature. Similarly, an author may have, say, a yacht and therefore need to be able to perform some of the operations which the captain performs. A consequence of all this is that much rehabilitation has to be geared to goals which people have in common. I wanted to point out that *some* vital goals are specific to individuals and that therefore *some* of the rehabilitation should be of an individual character.

I: I understand you on this point, and I suppose I must agree. Health can differ in character from one person to another. But now comes a question which I think must make you quail! How do we find out what a person's vital goals are? Also, when does a goal become vital, and just when does it cease to be vital? We had better devote a special conversation to this.

Thirteen

ON MINIMUM HAPPINESS

P: You rounded off our last conversation by raising what is probably the most fundamental question concerning my own and all other holistic theories of health. It is possible to formulate the question in different ways. The most everyday form is this: how are we to determine what a person should be able to do in order to count as healthy? Or, in my terminology: how are we to discover P's vital goals?

I: Yes, and you have shut the door on the possibility of just *asking* P. I mean, you have said that P may not be conscious of her or his vital goals.

P: Theoretically you are right. I cannot just content myself with asking P. Nevertheless, I have said that in the typical case P does know perfectly well what her or his vital goals are. In practice, it is often possible to go a long way by asking questions. But sometimes, as you have pointed out, we must have recourse to other sources of knowledge as well. The first consideration is what is to be the criterion of a goal's being vital. Having spent a great deal of time thrashing all this out in my mind, I eventually settled for what appeared to be the only feasible conceptual starting-point: the concept of happiness. A goal is a vital one if it is strongly related to the person's happiness. I spoke of it earlier as a prerequisite. A goal is vital for P, if it is a prerequisite for P's long-term happiness.

I: Why long-term?

P: Because the long-term perspective has value for us as human beings. If a goal has a major part in my happiness for no more than, say, a couple of hours, it has to count as extremely lightweight.

I: But when can we say that a goal is long-term? Is it a question of a week, a month, a year, or what? Or do you see it as a question of the rest of the person's life?

P: That is an impossible, even meaningless, question inasmuch as my concern is with conceptual analysis. Analysis of the concept of health will give us no precise knowledge on this point. However, we can make a stipulation. This is part of the operationalization which we have to undertake in health care and in research if we wish to have a reasonably precise concept of health. Personally, though, I don't think that the temporal dimension should play a large part in the general description of health. One reason is that the temporal dimension, too, has an individual aspect. Being happy during a determined but extremely short period can in itself be vital. It can, for example, be vital to an aged and frail grandmother to experience the happiness of attending a grandchild's wedding, which may be a matter of no more than an hour or two.

I: But let us consider the expression "minimum happiness" which is part of your definition. Why that? I don't even understand what minimum happiness *is*. I never see myself as being in a state of minimum happiness.

P: No, I hope that you are better off than that. I am not enamored of the expression "minimum happiness." It is one of those terminological constructions to which we sometimes must resort in order to satisfy requirements of precision. However, the following notion underlies the use of "minimum": health is not sufficient for complete happiness. You remember how I rejected the WHO definition in conversation five. To some extent, this is the same type of issue. Our concept of health is linked to our physical and mental abilities. But even if we have the best such abilities, and also make use of them, this cannot guarantee our being completely happy. Happiness depends also on favorable circumstances, and if we are to attain a high level of happiness we need to be uncommonly fortunate.

I: I sense that you have a rather wooden, almost primitive theory of happiness. Do you regard happiness as coming together of ability to act and favorable circumstances?

P: I may have a primitive theory of happiness. But your reaction is attributable to my being forced to express myself schematically. If we try to present things lucidly, it can seem wooden. The lucidity removes all poetic dimensions. This is especially evident when we speak of things which in many respects the poets have best described.

I: Then why not leave it completely to the poets to define happiness?

P: Now you're being impossible! I said that the poets in many respects have described happiness best. That does not mean that poets are better than, for instance, philosophers at elucidating the relationship between health and happiness.

I: Let us go back, then, to your picture of happiness. Is happiness a conglomerate of ability to act and favorable circumstances?

P: Not exactly. Happiness can be a consequence of ability to act, exercise of this ability, and favorable circumstances. But happiness can also be a consequence just of extremely favorable circumstances. One thing is that we can receive gifts. I am not thinking merely of concrete gifts but also of spiritual ones. For the Christian, for instance, happiness can come from religious awareness, from the awareness of Jesus as the savior of the world. Here we have a striking example of happiness which is not the result of any action on the person's part.

I: So being healthy is not the only way of achieving happiness.

P: No, our theory would be quite absurd if it led to such a result. Health is not identical with happiness. Nor is it a prerequisite for happiness. But health is a central factor in respect of happiness. Under normal circumstances, the healthy person can go a long way in her or his efforts to achieve happiness. However, we can to some extent compensate for lack of health. People with restricted ability to act can be helped by being offered special facilities for coping with their daily needs. This is a central aspect of rehabilitation. Occupational therapists see that the sick person's external circumstances are such as she or he can function in. Handicapped people can receive special aids; both home and work can be adapted. But these things are not the only ones that can bring happiness. I assume that happiness for you is not just a question of being able to function and work.

I: No, but I agree that the ability to act is basic to happiness. It strikes me that we need to talk more about the nature of happiness if your concept of health is to become completely plain.

Fourteen

ON THE NATURE OF HAPPINESS

P: I agree with you that we need to say more about happiness. I want to clear myself of the charge you have made that I have a wooden conception of happiness. We can start today's discussion from your having said that the ability to act is basic to happiness. You have an extremely famous predecessor who went even further in linking action and happiness. For Aristotle, happiness was identical with human activity which fulfilled designated conditions. Thus, Aristotle placed the essence of happiness, or the ontology of happiness, elsewhere than we usually do today. We normally conceive of happiness as a feeling, while for Aristotle it was no such thing.

I: Do you think we have reason to return to Aristotle's concept of happiness?

P: No, I don't. Above all, it would be a hopeless enterprise to attempt to reform a use of words which has become so firmly rooted as ours on this point. It is also firmly rooted in philosophy. One of the most influential ethical theories, utilitarianism (concerned with the greatest happiness of the greatest number), embraces a concept of happiness as a feeling or class of feelings. To change all such thinking would be impossible.

I: So have we anything to learn from Aristotle?

P: Yes, we have an immense amount to learn from Aristotle, and not just in this area. He was a foreruner in almost all areas of modern philosophy. In addition to his achievements in welfare theory and ethics, he laid the foundations of logic, semantics, and theory of science. But even if we limit ourselves to the question of the nature of happiness, our debt to him is great. Even if we do not share his view of happiness, we can quite easily interpret his observations in terms of our view. When Aristotle says, for instance, that moderate and thereby virtuous activity constitutes happiness, we can easily see that such activity is a prerequisite for lasting happiness in our sense.

What is more, Aristotle in effect assumed this too. The virtuous person also feels good in the sense of having pleasurable experiences.

I: Is that self-evident? Why should the moderate person be the one that always feels best? Sounds rather boring. Isn't the life rich in variation and excitement the happiest one?

P: I did not maintain that Aristotle's answer was adequate in response to our question as to what constitutes happiness. All I wish to say is that we have things to learn from him even if we read him principally from the point of view of what are the conditions for happiness. I think that a life which is varied and has a degree of excitement is a prerequisite for a high degree of happiness. But surely you would not call it the *basis* of happiness?

I: How do you mean?

P: How much happiness can we derive from variation and excitement if we lack basic security? Only when all of life's basic essentials are in good order can we start thinking of variation and excitement at all. What if you lost your job or even your family? What pleasure would you then get from an exciting life?

I: I agree it is a complicated business. So what is your prescription for happiness?

P: I have no well-thought-out prescription for happiness. But I think our discussion has gone off course. The brief digression on the subject of Aristotle has brought us to the question of a prescription for happiness. But our basic question was the conceptual one, "What is happiness?" The question, What are the paths to happiness?, can only partly be answered in philosophical discussion. Empirical psychological knowledge is indispensable. We must know a lot about a person's biological and psychological make-up if we are to say something tenable about the person's chances of happiness.

I: Excuse the digression. But what, then, is happiness for you?

P: I would like to summarize it in the following way. A person is happy if she or he finds existence to be on the whole as she or he wishes it to be. This includes external conditions in the form of economic circumstances and physical and cultural environment, and internal conditions in the form of energy, harmony, and emotional stability.

I: But wait a minute. That appears too simplified. You are saying that a person is automatically happy if both external and internal conditions are favorable. But life is not as simple as that. Happiness is a feeling over which we have no control. It can appear on the most surprising occasions. It may be more likely to appear when circumstances are favorable, but nothing is necessary about this.

P: You are making the discussion difficult and interesting. But first I must correct a mistake. I did not say quite what you think I said. A person is not happy because her or his situation is objectively favorable. Decisive is what the person wants, and whether the person judges the situation to be in accord with this.

I: But happiness is not just a question of a judgment and getting what we want. It's a feeling, sometimes an overwhelming feeling, which can come and go without our always understanding why.

P: I agree with you that a feeling exists that we call happiness. It can be strong—even, as you say, overwhelming. But I don't agree that happiness is entirely such a feeling.

I: You have lost me completely.

P: Consider this: you know a man that you would describe as a happy person. Or as having been happy during the past year, in contrast with previous years. You are thereby decribing him as having been happy during a long continuous period.

I: Where is the problem in that?

P: There is no problem for me, but possibly for you. Would you be ready to say that the man had experienced an overwhelming feeling of happiness non-stop during that entire year or more?

I: No, that sort of thing must be rare. But I suppose the feeling of happiness can sometimes be mild.

P: That's still not going far enough. What feeling of happiness do you think the man experiences when he is washing-up, or paying a parking fine, or for that matter sleeping?

I: I fear that you have slipped into a parody of philosophizing. You should take what I say as it is intended, not in such a dreadfully literal way. Obviously, I don't mean that the happy man feels happy 24 hours a day. What I mean is that the feeling of happiness is his dominant feeling during the period when I call him happy.

P: In saying that, though, you are getting close to my analysis. What I believe, you see, is that when we are happy we have a predisposition to experience a feeling of happiness. We tend to have this feeling, when we think of the source, the object, of our happiness, for instance when we think of a loved-one or of an approaching holiday. But we need not always have it.

I: So far I think I can agree. The main thing is that you don't forget the feeling of happiness.

P: I need to take up another point in your criticism. It is a point on which we do not agree. You implied that no connection exists between a person's judgment of the world around and the person's happiness. Happiness can come as a surprise, you said—it comes and goes without our always understanding why.

I: Yes, that is how it is.

P: On one level, I see what you mean. Yet it remains unsatisfactory because you have abandoned the hope of a thorough analysis of the concept of happiness.

I: Perhaps this is where we should stop analyzing.

P: I don't understand you at all! Stopping here would make theoretical welfare research impossible.

I: Sorry. Please go on.

P: A fine starting-point for understanding the nature of happiness is the assertion that happiness, like all other emotions, has an object. When we are happy, we are happy about something. Typical examples are being happy about success, about a gift, or about having discovered something desirable. We can describe in more general terms what such an object should be like. That is just what I was attempting to do when you interrupted me.

I: So you mean that the favorable circumstances are the object of happiness?

P: No, a person is happy about a situation which she or he finds is as she or he wishes it to be. The person's own judgment of the situation and the person's own wishes are decisive.

I: But do you mean, then, that the presence of such an object is sufficient, and that the person should be aware of it, for the person to be happy?

P: Yes, strictly speaking that is what I mean. But that the person has a feeling of happiness does not follow.

I: Is it, then, in your view, impossible to be happy about nothing at all, or to be surprised by a feeling of happiness?

P: Well, while I don't think we can be happy about nothing, it can happen that we are unaware of the object of our happiness. We can feel happy without knowing why. But that does not mean that the happiness has no object. Probably we would be able to discover the object if we tried, or someone might remind us. We can be surprised by a feeling of happiness in the sense that a favorable situation may arise quite unexpectedly. But happiness cannot be surprising in the sense of arising without having an object.

I: What kind of dogmatism is that? Is it impossible for a feeling of pleasure to arise without a change in the situation?

P: Ah, but you have introduced another concept altogether. Pleasure and happiness are completely different things. Sensations of pleasure do not have objects of the type that happiness has. For the most part, a distinctive characteristic of such sensations is that they are located in a specific part of the body. Pleasurable sensations of taste are located in the tongue, pleasurable sensations of sex are located in the genitals. But happiness cannot be localized in that way.

I: It is becoming increasingly plain to me how hedged about with qualifications your concept of happiness is. But I am glad that this *is* becoming plain. I can understand your theses more easily.

P: I am grateful that you press me so hard on this, because the concept of happiness has a central position in the theory of health I am attempting to introduce. No misunderstanding must exist between us concerning what happiness I have in mind.

I: Let me make a critical point of quite a different sort. Your concept of happiness is close to the concept of satisfaction with life. But to use satisfaction as the basis of welfare policy in general and of health policy in particular would be dangerous.

P: Are you thinking of the great differences in people's expectations?

I: Yes, some people have few and extremely modest expectations; therefore, they can be satisfied and happy in your sense in extremely modest material circumstances. Other people want and demand a great deal, and for that reason for them to be satisfied is difficult. What bothers me is whether we in some way are to feel a moral obligation to pay regard to the wishes and demands of these people.

P: I find hard to see why any such obligation should exist. In itself what you speak of is hardly a theoretical problem. A profound truth is that the person who has small expectations finds it easier to be happy than does the person who has great expectations. This a proper analysis of happiness must be able to express.

I: Well, let me try to express my misgivings better. My starting-point is your theory of health. For you, being in a state of good health is having the ablity to attain the vital goals. These goals are in turn linked with the person's minimum happiness. Take, now, the case of a person who has great expectations and who for this reason has a low degree of health. This person's level of minimum happiness is higher than that of many other people. Assume, now, that the duty of the health service is to assist people with a low degree of health. From this follows that people who expect a lot are likely to receive more attention than those who expect less. This appears to me to represent a large moral problem attaching to your theory.

P: Thank you for putting me in my place! Your argument deserves a long exposition in answer, but let me content myself with the following remarks. First, I have not once in our conversations up to now said a word about what is a reasonable distribution of the resources of health care. Second, I must point out that a shift of meaning has crept in: we have gone from speaking of people's wishes to speaking of their expectations and demands. In the latter part of your argument you speak of great expectations, of people who demand a lot. From considering the psychological aspect, we have shifted to consideration of the action aspect: from wishes we have shifted to demands, implying action. When you speak of a demanding person, we can easily imagine that we have before us a person vociferously demanding this or that in the way of health care. Only in this type of case, I would say, does the moral problem arise. We do not consider reasonable that the distribution of the resources of health care should follow the principle of "The louder you shout, the more you get." A person who is unhappy, because reality is far from being in accordance with her or his wishes, need not be a demanding person in the sense you mean. Third, what affects our health is whether we can attain our vital goals or not. But these goals constitute no more than a subdivision of what we must attain (which of our wishes we must have fulfilled) in order to be "perfectly" happy. Fourth (and this we spoke about before), health care can involve "goal care": the health of people can be improved through cautious influencing of their vital goals. Health care in such a case becomes a matter of helping a person adjust to reality. I regard such assistance as perfectly legitimate. I cannot see that the society has less warrant for helping a person who has difficulty in adjusting to reality than for helping a person with a fracture or pneumonia.

I: I content myself with that answer on this point. Now I will test another aspect of your theory. It bothers me that it has a static quality.

P: How do you mean?

I: Let us imagine a woman whose wishes are modest and few. What should she do in order to be happier? She should that all her wishes are fulfilled. But then what? Has she arrived at a point where she cannot possibly become happier? Has a ceiling been set by those wishes? If so, I find that absurd.

P: An illustration will help here. Imagine a young girl living in humble circumstances. Her parents are poor, and she expects little of life. What is more, the girl knows little about life outside the village where she lives. But her modest expectations are fulfilled, so in a way she is completely happy. However, we would be unreasonable to say that she is as happy as she is capable of being. She may have a host of slumbering talents. She may be a potential world chess champion or great musician. Will she not be happier if she discovers her talents and can develop them?

I: That is precisely my question. But you, not me, must answer it.

P: I shall do my best. First we must narrow the case a little. We can imagine instances where the girl would not be happier at all if she were to develop her talent for chess or for music. A world chess champion or renowned musician is constantly under public scrutiny, and we cannot be certain that the girl would be happier if she were to grow up to live under the strain of this. The simple life of the village may well be ideal for her.

I: Yes, the development of a talent can involve complications. We can be forced to lead a life which is full of stress and which for that reason impedes the fulfillment of our other wishes, ones having to do, for instance, with the integrity of the person or with peace and quiet. You are quite right about that. Having a lot of wishes to be fulfilled involves a greater risk of collision of interests.

P: So let us imagine that the level of ambition is lower. The young girl discovers that she has a talent both for chess and for music, and she then devotes herself to these things on an unpretentious level where she does not come into the public eye. Can we now reasonably suppose that in thus cultivating his two talents he is able to live a richer and happier life than otherwise?

I: Yes, that appears reasonable enough.

P: In light of examples such as this, I think the theory of happiness needs to be refined. In characterizing happiness we cannot just take present wishes into account; we must also consider such wishes as are potential, in other words, ones which would arise if the person became aware of other possibilities in life.

I: I understand and I agree. But I think we need to make further additions. You have said nothing about the significance of time. It came up in our discussion of vital goals, you remember. We can imagine a person who is happy during a brief period, say when spending a short time with a close friend. Are you prepared to call this person truly happy?

P: Yes, I don't think we should deny the person's happiness, though at the same time we should, as you imply, stress the temporary nature of it. Some happiness lasts only a short time, other happiness endures. The enduring kind is of most interest to us and for which we have to work. The vital goals of which I speak in my characterization of health are such as constitute the basis of enduring happiness.

I: What do you have to say, then, about the happiness which is based on a lie or on sheer self-deception? Think of a man who has a completely false image of himself and of the way people see him. Assume that this image is extremely favorable and that the harmony of the man's existence is based on it. Is that true happiness?

P: It is happiness, although a fragile happiness. At any moment the image can give way. Some meddlesome or malicious person can get into their head to undermine the man's self-image, probably with catastrophic result. Such happiness is built on sand, and I would not recommend self-deception to anyone seeking a happiness that will endure.

I: I agree with you. Now I feel a strong need to gather together the threads of our most recent conversations. Our discussion of happiness gave me a lot to think about, but at the same time I realized we were getting onto things not directly connected with the philosophy of health. We must have another look at the relationship between health and happiness. To me health is a prerequisite for long-term happiness. At the same time, I find equally obvious that health cannot guarantee happiness, for a lot of people are healthy but unhappy.

P: I am in complete agreement with you on that last point. Health is insufficient for happiness, even for minimum happiness as I have defined it. This we established in the conversation preceding this one.To identify health with happiness would be a grave impoverishment of the labguage. That would make impossible for us to say that any of the unhappy people in a war-torn country were healthy. On the other hand, I want to press you concerning your first point. You said that health is a prerequisite for happiness. Do you mean, without qualification, that no happiness can occur without health?

I: I am inclined to think that, but I have to think about it more. If we are seriously ill and incapable of seeing to ourselves and doing our work, it must be difficult for us to be happy. I suspect, incidentally, that a logical consequence of your concept of health is that a person who is in a state of ill-health cannot be happy. If we are unhealthy we cannot attain all our vital goals and thereby we lack the prerequisites for minimum happiness.

P: I am glad that you take my concept seriously. But you have neglected one or two things about it. In the first place, remember that both health and happiness are continuous dimensions. For a few minutes we have been speaking of them in a simplified way as if they were absolute states; we have forgotten the Galenic lesson. That a person's health is limited does not mean, naturally, that the person cannot attain a single vital goal. Only in the case of an extremely grave illness does the whole person suffer defeat, only in such a case are all vital goals more or less out of reach. But the main thing that needs to be said here is: though sick people cannot attain all their vital goals by their own efforts, nothing prevents these people attaining the goals with a little help. Health has to do with our own ability to attain vital goals, but the vital goals can also be attained with the aid of other people.

I: If all of a person's vital goals are such as to be attainable with the aid of other people, then it must be possible for a person in a state of grave ill-health to achieve at least minimum happiness. And to the extent that other wishes can be fulfilled with the assistance of other people, the person's happiness can be all the greater.

P: Precisely. In principle, nothing prevents ill-health being compatible with an extremely high degree of happiness.

I: But I think that most people's vital goals include being able to accomplish certain things, such as doing their housekeeping or working at a job. Other people's help is something for which an ill person should be grateful, but it is insufficient for the attainment of all of her or his vital goals.

P: You are perfectly right on that. For many people, having a particular ability to act is in itself a vital goal which cannot be fully compensated for. Therefore the lack of ability to act almost always has a negative effect on the person's happiness.

Fifteen

ON SUBJECTIVE HEALTH AND ILL-HEALTH

I: We have spent a long time on several of the theoretical foundations of your concept of health, especially its link with the concept of happiness. This has distracted my attention from one of the cornerstones of your concept of health which I would like to consider again. At bottom, health for you is ability to act. Being in a state of good health is being able to do certain things. As was said in our eigthth conversation, this idea only partly corresponds to our basic intuitions. Health, and even more so ill-health, also has so much to do with our experiences. We associate health with well-being, and we associate ill-health with pain and suffering. This is not directly evident in your conception. I remember your explaining this in conversation no 11. You have said, for instance, that sometimes ill-health does not involve suffering, and you have said that when suffering occurs it appears also in your system, because suffering always limits a person's ability to act.

P: Yet I see that you are still not satisfied.

I: No, I am not. I think a good theory of health must reflect the fundamental role which pain and suffering play in ill-health. The experience dimension must be much more plainly visible in the characterization of health and ill-health. It cannot just come in indirectly through the links with future happiness or unhappiness.

P: I see your point. I, too, think that the experience dimension should have an extremely prominent place in the description of health and ill-health. But I am not so convinced that it must have a prominent place in the definition of them. Let me clarify this.

Though we have attempted to be as concrete as possible (this with the aid of examples), our main concern has been abstract definition. We have sought to discover the distinctive characteristics of the general concept of health (also those of the general concept of ill-health). This is a fundamental task, but it is far from involving all that we can say about health and ill-health. The next step is to describe the different forms they can take. Especially when we deal with ill-health, this has already been done in great detail. In textbooks of pathology and medicine, we have access to detailed descriptions of the causes, forms of development, and consequences of morbidity. There we also find a lot about symptoms and external signs. Among the symptoms are often fatigue, pain, itching, and other forms of suffering. All this, then, is minutely described, and no risk exists of its being forgotten.

I: Now we have got onto the individual diseases. Earlier you have made a point of distinguishing between disease and ill-health. I think that generally feeling bad is also a feature of ill-health.

P: I would not deny that it is often that way. But definitions should as far as possible involve *universal* characteristics. My thesis has been that an inability to act is a better candidate for being a universal characteristic than is suffering.

I: Would it not be possible, then, to expand the definition to include other common characteristics too?

P: Yes, and I have already proposed it. I want to present this to you in more detail, first by introducing the notions of subjective health and ill-health, second by differentiating between disease, illness, and sickness. Though in everyday usage the words "disease," "illness," and "sickness" are not always kept rigidly apart, nuances are perceptible on which above all anthropologists and sociologists have focused in order to create reasonably lucid and distinct concepts. So in modern medical anthropology and sociology the three words designate three separate concepts.

I: What differences are involved? Is this a question of three different types of affliction?

P: No, not in the sense that tumors, infections, and thrombosis are three different types of affliction. It will become plainer to you, I think, if I say that the terms designate three different aspects of afflictions. The first aspect is the organic-biological. "Disease" designates the pathological change in the body. "Illness" designates the person's experience of this change. The term covers different sorts of negative feeling, including pain and other suffering about which we have talked so much. "Sickness," finally, designates the role people adopt once they have begun perceiving themselves as ill. An example of this is falling ill, perceiving this illness, and going to see a physician. Not that I am saying that we need to swallow this conceptual apparatus whole. I just wanted to draw your attention to the distinctions that are in use in research, as a background to the distinction I wish to make primarily between the term "disease" and the technical terms "illness" and "subjective ill-health." This last term concerns the person's own perception of her or his general state of health. What I propose is this: if you think you are in a state of ill-health, you are in a state of subjective ill-health.

I: I am not sure that we are on the right track. Such different reasons can arise for my thinking that I am ill. One reason, for instance, might be what the physician said.

P: Good, though now you are going a little too far ahead. I still wish to start from suffering, or from what we vaguely refer to as feeling ill. The paradigm case is the following. I wake up one morning and feel out of sorts. Perhaps I have a sore throat and feel tired and a bit low. I say to myself, "Oh, I don't feel well. I think I've got an infection." Thus, it can be said of me that on the basis of these experiences and sensations I judge that I have contracted an illness and that my health is affected. The least that can be said is that I am in a state of subjective ill-health.

I: But the probability is that you are objectively ill and that your health has declined.

P: Yes, naturally. No opposition whatsoever exists between subjective and objective ill-health. (Objective ill-health is what we have been referring to just as ill-health.) A person in a state of subjective ill-health is normally also in a state of objective ill-health. The person who is subjectively ill is normally also objectively ill.

I: Nevertheless, you would say that this is not always so?

P: Correct. Our discussion has already shown that a person can be in a state of objective ill-health without being in a state of subjective ill-health, an example being when the person is unconscious. Or the other way round: a person can be in a state of subjective ill-health without being in a state of objective ill-health. We can be mistaken about the limitedness of our ability to act. We can wake up believing that our health has declined because we perceive one or more physical symptom, but then we get up and get dressed and shake off our feeling of discomfort, and it turns out that we function as well as ever.

I: I agree that such cases occur. I myself have an odd tendency to feel a bit rough first thing in the morning. Nonetheless, I am normally soon up and about. I still think, though, that not everything has fallen into place in the expanded theory which we have before us. At least two things I find troublesome.

P: I'm all ears.

I: In the first place, you have in a previous conversation (no 11) emphasized that pain and suffering are connected with reduced ability to act. If we suffer greatly, you have said, this reduces our ability to act. So suffering, a form of subjective ill-health, automatically entails objective ill-health. That is my first observation.

P: One which is valuable and worth thinking about. And what is your second observation?

I: It has to do with the extent of the field of subjective ill-health. So much appears to be included—everything from sensations and other experiences to the judgment that we are in a state of ill-health.

P: It looks as though I could let you take over control of the discussion now. Your observations are extremely relevant. I have not yet offered a clear-cut definition of the concept of subjective ill-health. But I thought there might be a point in our discovering it together. A lot of the scholarly discussion on the subject of illness suffers from a similar lack of clarity.

I: What is your prescription for improving the situation?

P: First we need to draw the distinctions you have already indicated. We need to differentiate between sensations, emotions, moods, and judgments. I regard the judgment category—or cognitive category, to use the more technical term—as occupying a key position. For many purposes, we need a special concept for the case where a person judges or believes that she or he is in a state of ill-health or is ill.

I: Should we reserve the term "subjective ill-health" for this type of case, and introduce other terms for other cases?

P: That has something to be said for it. But we can wait until we have done more reflection. Note that the different sorts of experiences which I have in slovenly manner called feelings of illness are connected with the judgment that we are in a state of ill-health. Normally, the feelings are the reason for the judgment. So this warrants bringing the different categories together under one roof.

I: I can see the relation. But a person can have quite a different reason for the judgment than just feeling ill. I can have been told by my physician that my health is not what it should be, or I can have observed my body or behavior and drawn my own conclusions.

P: You are perfectly right. I can add a further complication. We can have sensations of pain and feelings of nausea and anxiety without for that reason making the judgment that anything is wrong with our health. Think of all those who do not have the concept of health in their mental repertoire—infants and persons with grave dementia, for instance. We can pursue an entirely parallel line of reasoning with regard to ill-health in the animal world. As far as we know, not even the most highly developed animal species have anything like our world of concepts, yet we do not doubt that animals, like infants and persons with dementia, can experience most forms of pain and suffering.

I: So what we are referring to as subjective ill-health appears as a complicated area.

P: Yes, though our brief conversation about it has carried the discussion forward. We need to distinguish these categories:

1. P's judgment that P is in a state of ill-health.
2. P's reasons for this judgment. These reasons can be a question of
    a. P's own sensations, emotions, and moods,
    b. P's own observations,
    c. other people's observations and diagnoses.
3. P's experiences of different sorts which do *not* form the basis for any judgment on the part of P but which for other people would constitute reasons for judging themselves to be in a state of ill-health.

I: Should we have different names for these categories?

P: Certainly, if we are to do accurate theorizing in this field. Otherwise it is sufficient to have the different possibilities in mind and be able to express them.

I: With that, I think, we can consider our investigation of the question of subjective health concluded.

Sixteen

CONSEQUENCES FOR HEALTH CARE

I: We have gone through some basic concepts which are central to your theory. Now the time has come to see of what use these concepts can be to us in day-to-day health care. In other words, we must see whether the theory can be put to use and appears reasonable. A question occurs to me. How can one show that a theory such as yours is reasonable? Does a test exist which we can use, such as we use in the case of empirical scientific theories?

P: No, we cannot test it in quite the same way. Unlike the majority of scientific theories, my theory of health does not concern causal relations between phenomena in the external world. It might be described as a characterization theory. It is intended to characterize a concept central in everyday life and, naturally, in health care. As I explained in Conversation Three, my aim is not just to describe how we use the term "health"—my aim is also in part normative. I am attempting to reconstruct the concept so as to make it usable in even the most thoroughgoing communication about health and related phenomena.

I: So we have no test at all?

P: But we do! All this dialogue of ours is an essential part of such a test. You have raised a number of central critical questions. In the first place, you have attempted to make the theory comprehensible. You have played a part in producing interpretations which have enabled us to consider more concretely the theory's field of application. By pointing out one or two absurd interpretations, you have forced me to explain myself more thoroughly or even modify the theory. You have contributed to ensuring that the theory is not self-contradictory, that it fulfills basic logical requirements. All this is part of the essential testing of theories concerning concepts.

I: Is the testing that we have done so far sufficient to establish the utility of your theory?

P: I don't think so. We have to remember, too, that utility can refer to different levels. First, we have the most general level of utility. It is the level on which WHO's concept of health operates. Here the question is of establishing the general direction in which health policy should go, with indication of the highest goals. This level should be acknowledged, but it does not have much to do with day-to-day health care. A more concrete but still general level is when the concept of health forms the basis of different operationalizations in the scientific or clinical context. I am thinking of things like the elaboration of instruments for the measurement of health and quality of life. Such instruments must be based on a conception of health, and we can see how differently a biostatistician and a holist would frame their instruments. But the most concrete use of the concept of health will occur if it can become a living factor in the daily work of health care.

I: How can it become that?

P: The concept of health becomes a living factor inasmuch as it forms the basis of the judgments of people's health which health-care professionals are performing all the time. In the biostatistical case, the judgment as to whether the person is healthy is based on whether an evident pathology exists. In the holistic case, the judgment is based on the person's general condition and on her or his ability or inability to act (as based on the general condition).

I: Have you any illusions about being able to reach that point with a philosophical theory of health?

P: No I have no such illusions. It would be absurd for me to believe that my particular variant of a theory should be able to have such an impact. Another thing is that if we take having an impact as implying implementation, a lot of other factors come into the picture. The time must be ripe. Influential institutions must be ready to invest in the theory. It must be easy to grasp. It must also be well marketed. But marketing is not within the philosopher's domain. The philosopher should not proselytize. The philosopher should do no more than argue the case, as dispassionately and objectively as possible, and the philosopher must listen to all counter-arguments and be ready to change her or his mind. Not that this means that the philosopher is to shun the practical. She or he must attempt to set out what it would involve, in concrete terms, to follow a particular ethical principle or embrace a particular concept. Therefore, I fully agree as to the relevance of considering the concrete applicability of the holistic theory of health.

I: But if you keep such an awfully low profile and will not proselytize, how can changes ever occur in the use of concepts? Haven't we the right to expect more of an effort on your part? Shouldn't you as philosopher play a more direct role in matters of health policy?

P: For me that is a difficult question; nevertheless, I hope I can give you a satisfactory answer. I would agree that we have a general ethical duty as private persons to take an active interest in politics. Society could not function if we did not at regular intervals take up positions concerning political questions. We should vote, and we should also—according to ability and disposition—become more concretely involved, for instance, by taking part in demonstrations. It is also reasonable that we as scholars sometimes take up political positions on questions concerning which we have expert knowledge. But I want to keep apart my role as a philosopher and my role as a (political) private person. Political activity is not dispassionate. If we have a deep political commitment to something, we may not be inclined to listen to all the counter-arguments. This at once comes into conflict with the philosophical, rational ideal. We have to test and evaluate all arguments according to their deserts.

I: I suppose I must accept the position you take on this. Perhaps it is up to us outsiders to evaluate philosophical proposals in the ethical and theoretical-conceptual fields. I suspect that's just what I am doing right now! But still I wish you were a more political person. However, let me now go back to our main line of reasoning and ask about the utility of the holistic concept. Would you say that your holistic concept of health is usable on all the three levels of which you spoke?

P: I think it can provide answers on all three levels. I have already given an indication of how. On the general level, the concept of health says where we can find the essence of health: in the ability to act and not in any particular state of the body. On the next level, this concept provides a guideline for the framing of instruments for measuring health. Then, on the most concrete level, it provides us with a starting-point for determining whether a person is healthy.

I: Do you think that your theory gives adequate answers on the concrete level? Can it draw a superfine dividing-line between the healthy part of the health scale and the sick?

P: I want to avoid speaking of the sick part of the scale. I prefer to think of that part of the scale as the ill-health part. My answer to your question is simple. The theory does not provide a basis for drawing a superfine line between health and ill-health. Nor, I think, does any rival theory. We can also question the desirability of such a fine line. Sometimes, for bureaucratic purposes, for instance, we require a reasonably sharp line. In such cases, we have to come to a joint decision as to where we shall draw it. One such case is what criteria have to be fulfilled in order that a person qualify for sickness benefits. In the clinical context, the decision should in most cases be arrived at, I consider, by way of a dialogue between clinician and patient. This will appear quite natural in light of a holistic theory. Patients themselves are almost always in the best position to see whether at a time they can attain their vital goals. So the patient is most often the person in a position to determine whether a situation is satisfactory. For this reason, it would be rather absurd for the clinician to attempt to find independent criteria for drawing a sharp line between health and ill-health in the case of the particular patient.

I: Is mental ill-health an exception to this?

P: Only some types of it. In psychoses when we have reason to say that the person's judgment has become limited, we can no longer depend upon what the person says about her or his vital goals or about when she or he has attained them. In such a case, the decision-making becomes more complicated. We may have to consult the patient's relatives, and we may have to judge on the basis of our knowledge of general psychology.

I: But surely people in perfect mental health can be wrong about their vital goals? What sort of answer do you think you would get if you asked a child, or for that matter a teenager, about her or his vital goals?

P: The point you are making is correct. Let me, then, clear up any misunderstanding by saying the following. The criterion of whether the person is healthy is not what the person says about her or his vital goals. What I am saying is that under normal circumstances the person's own words on the subject are the best means at our disposal for ascertaining the person's condition. But we must always take what is said with a pinch of salt and critically compare one statement with another. I: This would be a good time to leave this general methodological discussion and instead consider practical consequences of your theory and raise critical questions about it from a clinical point of view. In our next conversation I intend to ask you about the clinical diagnosis of disease and about judgments concerning health.

Seventeen

ON THE RELATION BETWEEN DISEASES AND HEALTH

I: We have spent a long time discussing health and ill-health, including their relation to happiness and vital goals. I should like now to bring the discussion down to the level of everyday life and the medical world with its diseases, diagnoses of diseases, and the treatment of diseases. Does your abstract reasoning affect this world of medicine at all? What do you think?

P: The everyday world of medicine contains much which my reasoning does not affect and which I have no reason to attempt to affect. What is more, I am ignorant of the largest part of what goes on in that world. The major consequence of an acceptance of our reasoning would be that the basic identification of disease would sometimes be different than it is at present.

I: How do you mean?

P: Well, the idea of biological and psychological abnormality, in the sense of unusualness, still sometimes governs the identification of new diseases. As you now know, I strongly dispute that idea. We have no reason to call unusual physical or mental functions diseases.

I: On what grounds, then, are we to identify diseases?

P: Diseases are such physical and mental processes as tend to lower the level of a person's health. Pneumonia, for instance, normally involves fever and fatigue and thereby reduces the person's ability to act. In the typical case, the person becomes bedridden and unable to work or do anything at all of importance. This, I would say, is the typical distinctive feature of a disease. A disease leads, in the typical case, to ill-health in the sense I have given this term.

I: Disease leads in the typical case to ill-health, you say. That implies that you think a person can have a disease but still be perfectly healthy. That sounds odd!

P: Is it so odd? I recently read an interesting article in one of the big daily papers about the diseases to which men are prone in old age. Prostate cancer was one of the diseases discussed. It often has an extremely slow course. Sometimes a man can harbor it for 20 years without perceiving any sign of it. To my delight, the paper put it this way: "Many elderly men have prostate cancer without its affecting their health." Evidently the journalist saw as perfectly possible to have a disease and yet be completely healthy.

I: I understand the idea even though I still find your manner of expressing it a little hard to take. But now I have a strong argument against your whole way of thinking about health. Almost all diseases have an initial phase during which the person is unaware of being afflicted. Even the most rapid and dangerous type of cancer starts with cellular changes which the person does not perceive. So during the time between onset and the emergence of visible signs and symptoms the person has a disease, even a serious one, yet is in perfect health in your sense of the term. That is absurd. Aren't you flying in the face of all normal linguistic usage?

P: You formulate your argument lucidly, and I am grateful to you for having put it forward. My answer, nevertheless, is that you are not arguing against my theory of health but against something else. Things become absurd only if I maintain that nothing should be done for the person who has just been diagnosed as having a serious disease. But even if the disease has not yet affected the person's health, it does constitute an extremely strong threat to her or his health, and such a threat should be dealt with by means of health care. Combating the disease thereby constitutes an essential element of preventive care in the case of this person.

I: What you are doing is putting partly new labels on well-known accepted phenomena. You leave reality as it is.

P: Good philosophical practice is to leave reality as it is. The philosopher's principal task is to map out and interpret, not to change. However, it is wrong to say that I completely leave reality as it is. If my analysis indicates that statements or practices are inconsistent or incorrect, I want to point this out and suggest a change.

I: What kind of practices or statements do you have in mind?

P: Sometimes it is essential that we call things by their proper names. Calling sexual deviation a disease was of drastic significance for a large group of people. The love life of these people was classed as sick and as requiring the intervention of society with medical means. Little thought lay behind this. Spitzer, whom I mentioned in one of our earlier conversations (no. 9), performed an analysis concerning homosexuality and came to the conclusion that homosexuality could not reasonably be classified as a disease on the grounds which had been adduced. I had come to the same conclusion. Similarly, I do not think that we should classify all other deviations from the normal as diseases. The way I see it, a cellular change or a change in an organ's way of functioning is not by definition a disease. Such changes or deviations should not be the object of health care unless we have good reason to infer that they do constitute a threat to health (understood holistically, as I advocate).

I: May I change track? Are there threats to health other than diseases?

P: Your question can be understood in two ways. First, it can be understood as concerning external threats. In that case, the question is infinitely broad and has no place in our discussion. It takes in, for instance, things which indirectly threaten health by causing disease—pollution, smoking, stress, a brutalized society, etc. But it I think you intend your question to be understood in the other way: as concerning internal threats, coming from the person's own body. In that case, my answer is simple. Injuries and defects are other typical causes of ill-health. Although a dividing-line is difficult to draw between these and diseases, a convention exists among medical theorists to speak of diseases as processes and to speak of injuries and defects as states. Injuries are taken to be acquired states, while defects are taken to be congenital. A number of American theorists employ the term "maladies" to cover all three—diseases, injuries, and defects.

I: But can ill-health come from anything other than maladies?

P: In my opinion, yes. Maladies are the typical causes of ill-health, but not the only ones. The ability of people to act can sharply decline because of being in the midst of one of life's crises or feeling existential anxiety—and thus they are not in good health.

I: But why not in that case call the crisis or the anxiety a disease or a malady?

P: Essentially, the answer is that neither a crisis nor a period of anxiety need lower the level of a person's health, or indeed, typically does so.

I: But if it were possible to pin down a type of crisis or anxiety which does generally lower the level of a person's health, would it not be a malady?

P: You have a logical point there. The way I define these things, every inner process or state which typically leads to a decline in health is a malady.

I: But all phenomena belong to categories. Inner processes which have a negative effect on health are no exception, and an inner process which thus affects health does so by virtue of belonging to a specific category. We must in principle be able to define this category, and it must always be a disease. We can reason in the same way about other maladies. So all causes of ill-health must be diseases or other maladies.

P: You have become a shrewd critical philosopher. Your basic argument is correct. Nonetheless, I must mention two qualifications which limit the validity of what you are saying. In the first place, you must remember that the conditions for A's health are not identical with those for B's health. This follows from the circumstance that A's vital goals are not identical with B's vital goals. Thus, it is perfectly feasible that a particular inner state of A limits A's ability to attain a goal which is vital to A but not to B. The identical state may exist in B, but its consequences are not the same. B's health does not suffer, because B does not have the vital goal. So the state is not a typical cause of ill-health, and therefore not a malady.

The second qualification is more pragmatic. Every state can be categorized in one way or the other. And if a state has a particular consequence C, it must have this consequence by virtue of belonging to a particular category. But at a particular historical moment we cannot find and define this category. Nor, therefore, can we give the category a name and add it to the nomenclature of diseases. Consequently, we can always say that we encounter instances of ill-health which are not attributable to known and named diseases.

I: When you speak of the relation between diseases and ill-health, I am struck by another thing I find troublesome. Your concept of health is individual-related. A person's health depends on what vital goals she or he has. Disease concepts, however, refer to general categories. Cancer can afflict anyone in the universe, and the different types of it have—at least in principle—universal definitions. But how can we have a functioning international language of disease if the conditions for people's health can vary?

P: The question is of major significance but also familiar to me. It means that the characterization of the concept of disease must be of such precision as to permit no misunderstanding. On the basic level, the answer is straightforward. We have a universal language of disease which is codified in the WHO nomenclatures and classifications. The reason we can have this universal language is that the majority of diseases tend to afflict people, and affect human activities, on a fundamental level. Many diseases involve pain and fatigue, thereby reducing the sick person's general ability to act. It does not, then, make much difference what actions are of importance to the person, and does not make much difference what the person's vital goals are. Thereby the disease is in almost all its occurrences a cause of reduced ability to act. This means that it fulfills our criteria of disease. However, we can consider a type of physical change which has a negative effect only on an extremely out-of-the-way function which is of importance only for special persons. My favorite example is that of a circus clown whose act requires that she wag her ears. Suppose that something goes wrong with a nerve and causes her to lose this ability. Her health, in my sense of the word, thereby suffers. But the type of change in question (assuming that it does not have other effects—causing pain, for instance) does not fulfill the criterion of lowering the level of people's health in general. So according to my theory, this neurological change does not qualify as a disease (or other malady).

I: Here we have a consequence confusing to common sense. I want to ask you the following question. Have you considered to what extent your classification of diseases coincides with the internationally accepted one?

P: It coincides extremely well. The majority of recognized diseases, injuries, and defects lead to some limitation of a person's ability to act, either directly or indirectly (this last through pain, fatigue, and other suffering). So they automatically go onto my list of maladies.

I: How odd that the two should coincide. In your opinion, then, does your definition in practice underlie the natural concept of disease?

P: Not my definition in the sense of something that I have formulated. But I believe my analysis follows the same line of thought as underlies the natural concept of disease.

I: Well, you won't be let off so easily. Eventually I will return to several more serious problems for your concept of health and for your concept of disease. They concern the relation of your theory of health to the sphere of handicap. First, though, I want to take up another burning question, to do with the work of health care and different ways of attaining health.

Eighteen

ON THE PATHS TO HEALTH

I: We have devoted quite a lot of time to examination of the nature of health in accordance with your holistic view of it. But I imagine most of us are more interested in the more practical question of how we attain and maintain health and avoid all threat to it. Not that this question is your concern as a philosopher. But do not some conclusions concerning paths to health follow from the definition itself. Let me explain. Health as you define it is a relation—you as a philosopher will call it a three-place relation—between a person's ablity to act, her or his vital goals, and a class of circumstances. We can influence this relation in, basically, three ways. First, we can influence the ability to act. This is normally the relevant form of influence in health care and also in health promotion. We influence people's bodies when we cure or alleviate illnesses; we try also to indirectly influence the body when we make recommendations about diet and lifestyle. The way you see things, however, this cannot be the only way of influencing a person's health: equally feasible is to influence it through alteration of the person's vital goals or circumstances.

P: That's a correct conclusion. In principle, we can affect health by influencing any of the three elements of the relation.

I: But that produces an odd state of affairs. You imply that I could improve my health by asking less of life and thereby bringing about a better correspondence between my expectations and my abilities. Or that I could improve my health through a change of environment, in this case bringing about a better correspondence between my circumstances and my abilities.These appear as a marvellous answer to the economic crisis of health care. What a dramatic reduction of costs they would mean! Have you spoken to any local politicians about these conclusions?

P: No, I have not been wooing any politicians. I was expecting your question—you are not the first to have asked it. I have quite a lot to say in answer to it. And when you have heard what I have to say, you will see why I can look forward to no enthusiasm on the part of the local, national or even international politicians.

You spoke first of being able to improve your health just by lowering your expectations and thus producing a better balance between your expectations and your abilities. You could thereby influence your health through a simple decision. You find that quite absurd. I agree that it sounds absurd. Note, however, that you have chosen another concept than the one which I use in my theory and which we have discussed thoroughly in our conversations. You speak of expectations, but expectations and vital goals are not the same thing. We can change an expectation through a simple decision, but we cannot change a vital goal so easily.

I: Can we not change our vital goals ourselves?

P: Vital goals can be changed, through the influence of other people or through the passage of time. Making the relevant decision can also help to change a vital goal, provided the time is ripe. But remember the definition: a vital goal is a state of affairs which is a prerequisite for the person's long-term happiness. Not all of the goals we set ourselves are vital goals in this sense. A genuine change of a vital goal requires a change of personality; it requires that we genuinely set a high value on the new state of affairs promised by the change of vital goal.

I: Are we to regard psychotherapy as health care which has to do with the vital goals? Is psychotherapy a sort of "goal care"?

P: Yes, to some extent. Psychotherapy is assuredly the traditional form of treatment which more than any other deals with questions concerning a person's goals in life and the person's relation to these goals. Psychotherapy has the task of enabling the person to achieve greater self-knowledge and thereby a realistic attitude to life's demands. Realism is a prerequisite for health. But to speak of psychotherapy as being concerned only with "goal care"would be wrong. It is equally concerned with psychological factors which limit a person's ability to act.

I: If I understand you rightly, though, we can improve our health by lowering the level of ambition of our vital goals. When I got onto this before, your objection came to nothing more than that changing vital goals is difficult. We cannot just change vital goals by writ. A more consistent application of "goal care" would not necessarily reduce health-care costs. But is "goal care" even feasible? I see two difficulties with it. First, I wonder whether the lowering of goals *does* cause an improvement in health. I still have trouble on the conceptual level, you see. Second, is an attempt to influence people's vital goals ethical? Should we not leave people's personalities alone?

P: Splendid! Let us first consider your remaining conceptual misgivings. What prevents you from believing that a person's health can be improved through a change in vital goals? The best approach is to think of types of mental ill-health which are associated with a defective apprehension of reality and a lack of realism. The people afflicted think themselves more capable than they are. For this reason, a person's goals, including some of the vital ones, can be totally unrealistic. But the person's ability to act can be as good as that of an ordinary "healthy" person. In a case like this, the work of health care should focus on the unrealistic apprehension of reality and the unrealistic goals. Then about the ethics of it. You are correct that health-care staff must not indiscriminately influence a person's vital goals. However, this is just a special case of the need always to respect the patient's integrity and right of self-determination. This need applies equally to the psychotherapeutic treatment about which we are speaking with reference to "goal care." Caution and respect are likely to be of greater importance than ever when we consider things bound up with the patient's personality.

I: Then let us leave "goal care" and instead consider "environmental care." Following your three-place relation, we can also improve our health by changing our environment (our circumstances), or by adapting it to our ability. A person unhealthy in environment 1 might not be unhealthy in environment 2. This is another thing that does not appear in accord with our ordinary notion of the nature of health.

P: Here too, with regard to this element of the relation, I would ask you to carefully consider how I have expressed myself. I have not said that people are healthy if they can attain their vital goals in their present circumstances. That would have absurd consequences. It would mean that we could change people's health just by putting them in different environments. A person put in prison, for instance, would immediately become much less healthy, since imprisonment severely restricts a person's ability to act. Conversely, a person put in a position of power would at once become healthier.

I: But *is* that so absurd? Don't you think the prison environment would indeed lead to ill-health? And don't you think the freedom given by being in a position of power would indeed lead to better health?

P: Yes, in the long run these situations can lead to changes in health. But my argument was not about that sort of imaginable causal relation. The theory of health you appear to be ascribing to me, but which is not mine, would have as an automatic consequence that the prisoner was in poor health and the powerful person in exceedingly good health. The prisoner would be in poor health because the walls and the guards would prevent performance of many activities of everyday life. The person put in a position of power would be in exceedingly good health because this position would make possible the performance of actions beyond the range of most of us.

I: But what precisely are you saying about the circumstances that are an element of the definition of health?

P: I said that a healthy person must have the ability to attain her or his vital goals under reasonable circumstances. I am glad to have the opportunity to elaborate on this now. The word "reasonable" indicates the way in which health is an evaluative concept. If I ascribe health to another person (or to myself), I am indicating that the person can attain her or his vital goals under circumstances which I evaluate as being reasonable. Such evaluations can vary. If a person makes a statement like, "Harry's healthy," but says nothing about the grounds for this statement, then the statement is unclear and thereby inadequate for either clinical or scientific communication. Therefore, we need to know what shall be considered reasonable circumstances. We must have agreement about a set or class of standard circumstances. The majority of such agreements, I believe, are not set down in print but just implied. Communication among the members of a team of carers over a long period can in the end establish what they see as the class of standard circumstances.

I: Wouldn't those circumstances usually be called (if I may provoke you a little) *normal* ones?

P: Yes, I am sure you are right. But in this context the word "normal" would not say more than my word "standard." What is more, the word "normal" would be confusing rather than anything else, if it were associated with the idea of frequency and usualness.

I: But shouldn't the standard circumstances be the circumstances which the person most frequently encounters?

P: A link does exist between them. A concept of standard circumstances which only covered extremely unusual situations would be unusable. We must not identify standard circumstances with the most commonly occurring circumstances. That would land us with problems like those we discussed in one of our earlier conversations (Six). Circumstances do exist which are extremely unusual but which are such that we find reasonable that a person should be able to cope with them. In other words, some unusual circumstances should be included in the class of standard circumstances. A winter temperature of 20 °C is highly unusual in Sweden, but it is a temperature which we are expected to cope with if we are healthy.

I: I can see that if we are to continue our reasoning on this point we must first analyze how we are to identify different types of circumstances. I agree with you about the unusualness of the winter temperature of 20, but such a temperature is not unusual in general (not unusual in summer). It depends upon the level of generalization.

P: The different levels of generalization play a key role when we come to identifying types of circumstances. The frequency of the types can be completely dependent on it. But we encounter consequences in both directions. If we go to sufficient length in specifying situations, no situation is usual; if we go to sufficient length in generalization, every situation is usual. Therefore, we have to decide on a level of generalization which differentiates, so that some types of situation are usual while others are unusual. But suppose that we have done this and concentrate on the unusual situations. Let us say they occur once a year. I maintain that among these situations will be ones that we should include among the standard situations relevant to the definition of health. Some unusual situations are such that the healthy person should be able to act within them.

I: So can we influence a person's health by changing the person's environment? We cannot do it just by sending the person to prison or on holiday to Spain.

P: Consider the way you express yourself. We easily get the impression that you are speaking of a causal relation. We can exercise a causal influence on a person by moving her or him to another environment. A person's mental health must be worse after a long term of imprisonment. In the other direction, illnesses can be cured or alleviated through a change of environment, the person's health being thereby improved. But my definition of health does not refer to this. So we need to formulate the question as follows. Can we automatically change the state of a person's health by influencing the circumstances-element of the health relation? The answer is that yes, we can—but only if we alter the entire class of standard circumstances.

I: It sounds like a dreadfully complicated operation. I have no idea how we can accomplish it.

P: From the practical point of view, the operation need not be complicated at all. The easiest way is to have the person move permanently to a different environment. Imagine that a person from Sweden moves permanently to the Central African Republic. It involves a radical change in living conditions.

Both climate and culture are completely different from what the person has been used to. What count as reasonable living conditions in the Central African Republic are different from what count as reasonable living conditions in Sweden. Reasonable living conditions in the Central African Republic include, for instance, a temperature and a level of humidity which are consistently extremely high. Though the environment may be free from the Western tempo, other causes of stress exist—insecurity in relation to the authorities, above all, but also insecurity in the face of natural threats such as disease and wild animals. If the person can attain her or his vital goals in Sweden but not in the Central African Republic, so the person is healthy in the original environment but not in the new one.

I: When I hear you speak about reasonable living conditions I find myself wondering once more who decides what is reasonable and what is not. Is human health ultimately a question of power? Do authorities dictate what is reasonable? All this has a nasty taste of the sort of thing that came up in our discussion of mental health in the USSR. What is the point of all our meticulous analysis if, when all is said and done, the decision-making is in the hands of politicians and civil servants?

P: Now you have me confused. You hop over to a completely different issue—a political one—and wonder how conceptual analyses of this type can influence political decisions and their application in the field of health care. This is an interesting issue in its way, but the discussion of it would require quite a different approach and quite different knowledge on our part.

I: I cannot see that my point is illegitimate. Do you want to put the lid on the discussion?

P: No, that is the last thing I want to do. But I want our conversation to be coherent and fruitful. Imagine if our discussion had been about the concept of knowledge instead of about the concept of health. Imagine if we had in Socratic fashion been considering how best to characterize the concept of knowledge. Imagine that then you had suddenly started talking about the school and the power of teachers to define what is knowledge and what is relevant knowledge. I imagine that you had complained that with our abstract discussion we would not be able to influence the school's view of knowledge and the manner of conveying knowledge. Wouldn't that mean we were getting sidetracked?

I: You are being unfair. I was reacting to an extremely specific point in your characterization of health, namely, the idea of relating health to a class of reasonable or standard circumstances. Something is abitrary in that idea—it invites those in power to decide this or that as the fancy takes them. Your concept of health contains a dangerous element.

P: Now I think that you are the one being unfair. My analysis has led me to see the concept of health as primarily an evaluative concept, the evaluation being found principally in the element of "reasonable circumstances." As in the case of all other evaluative concepts, especially the purely evaluative ones like "good," "bad," "beautiful," and "ugly," people will use it in different ways, ascribe different objects to it, in line with their personal values. This means that authorities can use the concept of health in their own way, and can by virtue of their power exercise a great influence upon how it is used. Yes, authorities can steamroller their own values through, but does that make evaluative concepts generally dangerous ones?

I: Who, though, is to evaluate the circumstances, the way you see it?

P: It depends on what we are concerned with. Everyday occasions exist where anyone at all can do it. If you're going to say something about your health, or I'm going to say something about mine, or either of us is going to say something about the other's, we are free to make our appraisals concerning what circumstances are reasonable. If, however, health-care staff are involved in making appraisals which will constitute the basis for care, the situation is different. The appraisals need to be reasonably uniform. A legal rights dimension comes to the fore here. Not that the uniform appraisal need be a matter of a dictatorial resolution. At least in a democratic country, it is reasonable to discuss a major matter such as this and arrive at a consensus.

I: So you think, do you, that we should be discussing in order to reach a consensus concerning what are "standard circumstances" in connection with decision-making in health care? Isn't something ridiculous about spending time on that?

P: Be careful not to prematurely condemn something as ridiculous. I have not yet said anything about the degree of urgency of such discussion. Probably it won't have the highest priority in times of economic cut-back. All I am saying is that discussion and decision concerning the class of standard circumstances is a prerequisite for speaking about health with reasonable precision. Sometimes precision is not imperative, and then we can assuredly wait. Sometimes, however, precision is imperative.

I: Such as when?

P: Think of the sphere of rehabilitation. A physical therapist, occupational therapist, or doctor of rehabilitative medicine, seeking to improve the health of people through directly influencing their ability to act, must have an exact language. In order to be able to judge the progress of patients at all, rehabilitation staff must constantly define and redefine factors to do with circumstances. Take, for instance, a person who has been paralyzed in one leg and is learning to walk again. First, the person takes a few stumbling steps on an even floor and with the direct support of staff members. Then the person can cross a room without support. Only after this can the person walk—with difficulty—on uneven ground. The next stage is to go up a few steps. Next the person does different things during the same stage of training. And so it goes on. At each stage we must specify under what circumstances the person can walk. Without this, we cannot properly comprehend what progress the person has made. Nor, without this, can proper communication occur between one member of the care team and another, or between a member of the care team and other staff.

I: All of that I understand, and I agree with you about it. But up to now you have spoken of different degrees of ill-health. The circumstances have been limited. From this does not follow, that you need to describe the circumstances in the ideal state of health.

P: Not in the ideal state of health. But think again of the latitudes of health. We must differentiate between perfect health and such states as are no more than acceptable. Normally, when we give a person a clean bill of health we have to be content with the acceptable. We must know when we have reason to say that a person is rehabilitated.

I: In that case, have we any interest in finding a general set of standard circumstances? Should we not just consider the circumstances the person can encounter? Ought not this to follow from your individual-related concept of health?

P: I don't mean that in the situation we should attempt to ascertain all conceivable circumstances. We can restrict ourselves to what can reasonably occur. Remember, though, that sometimes completely unpredictable things do occur which can restrict the ability of people to act. The use of the general expression "standard circumstances" is designed to exclude such things.

I: The time has come to finish our discussion of the paths to health. In the next conversation I want to take up the question of handicap and health.

Nineteen

ON HEALTH AND HANDICAP

I: Though I am not satisfied with all of your answers, I must admit that you have provided answers to quite a lot of my original questions. Now, I intend to focus on an area which must be troublesome for you. I am thinking of the relation of your theory to handicap and impairment. I find hard to see how you can establish a dividing-line between ill-health and handicap. Your concept of ill-health is based on the idea of inability to act, and such inability is the core of the concepts of disability and handicap. The person with a handicap or disability lacks an ability in some area. But isn't it absurd to equate handicap with disease or illness? The organizations for the handicapped resist just such an equation.

P: Thank you for pressing me on this point. My theory entails a strong relationship between ill-health and handicap. At first sight this appears odd, and troublesome for the theory, but it may not continue to appear so once I have explained myself. Let me first say something about the terminology. "Handicap" is not the only term employed to designate a hindrance to action. The term is becoming increasingly rare in scientific and treatment contexts. It is considered to have a negative ring and is therefore most often replaced with "disablement." Furthermore, we now have a complicated nomenclature, originally drawn up by WHO, which distinguishes three main levels of disablement: impairment, disability, and handicap. The first level concerns organ disorders and the immediate effects of such disorders. The second level has to do principally with motor and behavioral disorders. The third level, that of handicap, covers such disablement as arises through contact with the circumstances and demands of the world around. All this is an interesting area worth going into more deeply. At the same time, to draw sharp lines between the different conceptual levels appears to be difficult.

However, in this discussion I will not go into these subtleties. I take the liberty of using the word "handicap" to refer to all types of disablement, from the inability to move a finger to the inability to manage funds or perform a job. I depend upon you to take the word as having no negative ring whatsoever. The relation to my concept of ill-health is equally evident over the whole field.

I: Does any difference occur, then, between being in a state of ill-health and having a handicap? Any difference, for example, concerning circumstances or goals?

P: Here we have to watch our words. Are we speaking about my concept of ill health and my concept of handicap? Or are we comparing my concept of ill-health with some official concept of handicap?

I: *Is* there an official concept of handicap?

P: Yes indeed. In the WHO nomenclature, handicap receives the following definition: "In the context of health experience, a handicap is a disadvantage for a given individual, resulting from an impairment or a disability, that limits or prevents the fulfilment of a role that is normal (depending on age, sex, and social and cultural factors) for that individual."

I: I note several things. First, we have that everlasting normality again. I wonder what sort of normality it is. Is it a pure frequency normality? Then, I see that WHO has introduced the concept of role. I wonder how it relates to your goals. And then I see that a handicap, according to WHO, must derive from an impairment or a disability. This means that it must derive from a more basic disorder within our own person. According to WHO, then, a purely external hindrance cannot produce handicap. Ordinary language says differently: in everyday life you can be handicapped through someone's putting a hindrance directly in your way.

P: That is an interesting observation. But it also means that WHO's concept of handicap is closer to my concept of ill-health than is the broader everyday concept of handicap. WHO is also, naturally, proposing a nomenclature designed to help health-care staff to classify impairments and handicaps.

I: But what about the concept of role?

P: Role is a concept of which I have not hitherto taken account in our discussion. This is not to say, though, that I think it of minor significance. It is of great sociological importance, not least medico-sociological. We can speak of adopting a sick-role.

I: What is a role in this context?

P: I see a role as a fixed set of normative relations. When we adopt an occupational role, it involves acceptance of the norms which prevail in the occupation. If I take on the role of policeofficer, for instance, I am accepting the norms which prevail in police work: I am accepting the duties, powers, and prohibitions associated with being a policeofficer. These norms can be codified or tacit, being in this last case conventions which—despite not being in writing—in practice govern the work in the particular occupational culture.

I: Role appears to be a concept completely different from your concept of goals. How can we compare the two?

P: Even though they are quite different, we can easily make a comparison. Why? Well, because the concept of role is used in the WHO manual to indicate the goals which are relevant to a person's actions. A person is considered to have a handicap if unable to fulfill the role which is normal for her or his age, sex, etc.

I: So role must be an immensely broad concept. It cannot just be a question of an occupational role.

P: Quite right. The occupational role is no more than a small part of it all—nor does everyone have such a role. What WHO must have in mind here is a general societal role, which for many people includes an occupational role. It is not altogether easy to see what this societal role is to involve. All we can say is that it must involve less than what I mean by vital goals.

I: Why?

P: Not all that we want to do is included in one or other of our roles in life. We can have vital goals outside our household role, our occupational role, our citizen role, or indeed any other role of which we can think. Fortunately, not everything in our lives is regulated.

I: So you find the WHO definition too limited?

P: Yes—and I am not alone in that. Several other critics, including people working in health care, would like to see the WHO definition altered.

I: But if we define handicap in your terms—in terms of vital goals—the similarity between ill-health and handicap becomes even greater. Does any difference exist at all? Don't handicap and ill-health become identical?

P: Yes, that is a consequence: having a handicap and being in a state of ill-health become the same.

I: Let me then apply your own critical method, that which you used concerning WHO's concept of health. For the normal user of the terms "ill-health" and "handicap," the terms refer to two distinct concepts, and a loss of conceptual richness occurs if we use them to refer to the same thing.

P: I have spent a lot of time thinking about that and I have also felt uneasy about it. Nonetheless, I can defend myself. A first point, again, is that we must not identify ill-health with disease or illness. National representatives of the disabled have emphasized that handicaps are not diseases or illnesses. On this point, we can be totally in agreement. Handicaps—variants of ill-health—are not physical processes or physical injuries, but they can result from such processes or injuries. This, incidentally, is unmistakably built into WHO's view of handicaps as consequences of impairments or disabilities.

I: But you can also conceive of other causes of ill-health. Should we have the same openness when we consider the causes of handicap?

P: I don't see why not. I think this is just as reasonable for the explication of ill-health as for the explication of handicap. Assuredly, a handicap could derive from, for instance, an existential crisis.

I: But you still see handicap, just like ill-health, as having internal causes. The causes are to be found within the person's body or mind.

P: That is a tricky question. It concerns what the purpose of my analysis should be. Up to now, my purpose has been to reason about, and suggest how to improve, the concept of handicap drawn up by WHO for use in health care. For such a purpose, it appears reasonable to take as a premise that the inability which constitutes a handicap has an internal cause. Thus, the handicaps which are the subject of health care will scarcely be directly due to, for instance, a war, a strike, or a political decision, nor directly due to more individual external hindrances, such as difficulty in finding a place to live or a job.

I: But you think a more general concept of handicap exists which can include such things?

P: Yes, indeed. Even the everyday concept has that openness. We are handicapped by being in a handicapping situation. How this situation has arisen is of no significance for the use of the term. It may have arisen through an internal hindrance (for instance disease), through an external one, or through the fact that demands—and thereby the person's goals of action—have been set too high.

I: I have been told that representatives of the disabled have expressed the wish that this open concept should find general acceptance in health care?

P: Yes, they prefer to speak of handicapping situations, one reason for this being to avoid putting the handicap label on the individual.

I: Obviously, the open concept of handicap is different from your concept of ill-health. Ill-health for you is inability to act. But this inability derives from within.

P: Quite right.

I: But let me return to my basic critical question concerning handicaps which do have internal causes but which would not normally be classed as instances of ill-health. I am thinking especially of learning-disabilities. A person gravely learning-disabled is quite incapable of achieving her or his vital goals. The person cannot handle the most basic requirements of living. Here in Sweden she or he receives special care, not the ordinary health care. To say that such a person has a reduced level of health does not seem reasonable.

P: What you say is basically correct. The gravely learning-disabled person has a severely reduced level of health according to my system. Normally, however, this ill-health is due not to a disease but to a congenital defect. Sometimes we know what this defect is, as in the case of Down's syndrome, and we can give a full description of it on the chromosome level. The learning-disabled are normally also taken care of at special institutions and not at ordinary hospitals. But institutionalization and bureaucratic classification can scarcely, I think, be of importance for our conceptual analysis. Also in the old days, special institutions for special types of disease existed. For a long time we have had, and still have (even if to a lesser extent), special institutions for the mentally ill. Tuberculosis and epilepsy have been treated in special nursing-homes. Care of long-term patients is now for the most part separate from ordinary hospital care. When we consider institutional arrangements, the decisive consideration must be what is suitable. This is a question of finding the most rational and most humane way of treating people with different sorts of problems.

I: I have no wish to make a big issue of the institution question. My feeling for the language is decisive for me. I would never call a learning-disabled person ill.

P: There we are again. You mix up ill-health and disease. I said just now that a learning-disability can come from a defect but seldom comes from a disease in my sense of the word. But I have no wish to quibble about this. You would probably not say, either, that a person who has a learning-disability is in a state of ill-health. A normal use of the term ill-health balks at such defects. I am well aware of this and realize that here my analysis has a stipulative effect. You remember what I said about explications of concepts in one of our earliest conversations (no. 3).

I: So here is a case where we have to accept that the system has an effect which is in conflict with our intuitions.

P: That need not by any means be a bad thing. If our previous use of terms was unsystematic, perhaps we now have a more efficient instrument both for theorizing and for practice.

I: You don't think that we are reintroducing the risk, which we were talking about just now, of the labeling of handicapped people?

P: No, I don't think we are. Let us suppose that we adopt the basic relational view of handicap that we have just been discussing. This means that we see a handicap as deriving from a situation. All the elements of a situation—characteristics of the person, characteristics of the demands or goals which have been put forward, characteristics of the human and non-human environment—can be the "origin" of the handicap. The handicap consists in the interplay between the terms of the relation. But in the analysis of a handicapping situation, we may discover that one of the terms plays an especially large part in producing the handicap. We "blame" this term and direct our therapy toward it. I find that a perfectly rational procedure.

We must sometimes, though assuredly less often than has traditionally been the case, assert that a person's inability to perform complicated actions derives chiefly from a characteristic of the person. We then speak of the person as having a learning-disability and give her or him the relevant help and support, sometimes within the framework of an institution. Through this we have in a sense located the handicap within the person. If any labeling has occurred, it has taken place here. I find hard to see that a characterization in terms of ill-health would make a decisive difference in this respect.

I: Though I am not completely satisfied, I think the time has come to finish this difficult conversation. In our next conversation, I want to take up the question of need for care, and of comparison between different types of such need.

Twenty

ON PRIORITIZATION AND NEED FOR CARE

I: I thought today we might take up the highly topical question of prioritization in health care. You know of the report of the Swedish Parliamentary Priorities Commission, *Priorities in Health Care*, which is being widely discussed. To hear what a philosopher of health has to contribute to the discussion would be interesting. Can your analysis of health offer us guidance when we have to decide which states of ill-health should receive priority?

P: I think the analysis can be of great help in ascertaining where the problems are, and also in finding some of the solutions. Yet I am well aware that much analysis remains to be done. But come on! We can see where the discussion takes us.

I: The Commission's analysis is based on three main ethical principles: (1) the principle of human dignity, (2) the principle of need or solidarity, and (3) the principle of cost-effectiveness. The first of these principles rates highest and has the highest priority, the third rates lowest and has the lowest priority. The import of the principle of human dignity in this context is that all people, regardless of race, sex, age or lifestyle, have an equal right to health care. The principle of need asserts here that the degree of the person's need for care shall determine whether she or he should receive it—and if so, how much. The cost-efficiency principle, finally, dictates that cost and efficiency shall be weighty factors when it comes to choosing one or another form of treatment.

P: A lot of questions arise even when we consider the three principles themselves. We can ask about the deeper ethical reasoning behind the particular choice, and we can ask how the principles are to be weighed against one another in practice. But I think that for today we should forget these questions. In the first place I am in sympathy with the principles and also with the order of priority. In the second place, our dialogue as a whole is not about ethical principles but about health concepts. So I find more reasonable that we should expend our energy on the closer interpretation of just one of the three principles.

I: How do you mean?

P: I mean that we should thoroughly analyze the second principle, the need or solidarity principle. It is in practice the key principle, and it greatly involves the concepts we have been discussing in all our conversations.

I: How can the second principle be the key one? I said just now that the principle of human dignity rates highest.

P: Yes, but the principle of human dignity offers no more than the ethical foundation for prioritization. It guarantees in the first place that priority will not be based on race, sex, status, age, or lifestyle. Beyond that, however, this principle can offer us little guidance. The more difficult work of prioritization starts with the principle of need. That's why this principle is in practice the key one.

I: I understand. But how can you say that the principle of need uses the concepts we have discussed? Up to now we haven't said a word about need for care.

P: No, we haven't. But I'm going to explain what I mean.

I: First, we need to know what the principle of need or solidarity involves. Let me read from the report. "Resources should be committed to those fields (activities, individuals) where needs are greatest." It also says (p.105): "The principle of need and solidarity means giving most of the resources of care to those who are most in need, if priorities have to be decided among effective measures. This applies even if, as a consequence, not everybody's needs will be provided for. The reason for choosing this principle is that it is closely related to the fundamental humanitarian reason for care." Solidarity receives the following interpretation (p. 105): "Solidarity means not only equal opportunities of care but also of an effort to equalise the outcome of care as far as possible. One desirable goal of care is for everybody to achieve the best possible health and quality of life.... Solidarity also means devoting special consideration to the needs of the weakest, e.g. those who are not aware of their human dignity, those who have less chance than others of making their voices heard or exercising their rights."

P: Good. Now we have some starting-points for our analysis. We can especially bear in mind the idea that care shall in the first place go to those most in need, those with the most serious diseases and the worst quality of life. Let's focus on the concept of need for a while. "Need" is a word that designates a relation. A person P has a need for an object O in order to attain a goal G. Here is an example. A carpenter needs a hammer to put nails in. To speak just of having a need, period, is meaningless. Unfortunately, this basic fact has been forgotten in parts of the literature. Sometimes an author will speak of people's needs without attempting to determine the goals to which the needs are related.

I: But how does the need for care fit into this? What is the object of the need and what is its goal?

P: The object is straightforward enough. A person in need of care has a need for care, so care is the object. The goal, though, is not so self-evident. Here we see one of the merits of the report. It attempts to indicate, with some clarity, the goal of the need for care. It does so with the aid of two concepts: health, and quality of life. This is what we read on p. 99: "The Commission will use the term 'need' in the sense of health-related and quality-of-life-related need. This implies that, the more severe one's illness or injury, or the poorer one's quality of life as a result, the greater one's need will be." Further, the concept of health and the concept of quality of life shall be regarded as having the same weight when it comes to estimating need for care. On p. 26 we read: "In the event of illness and injury, needs relating to the quality of life carry as much weight as those relating to health."

I: Is that not problematic? Can quality of life in general be a need-related goal in health care? Does it not involve unmanageable consequences? Can health care live up to such a goal?

P: Why do you see this goal as so problematic?

I: If the task of health-care staff should be to devote themselves to improving the general quality of life of patients, where would the task ever end? Imagine if physicians and nurses were to busy themselves with the marital difficulties, unemployment, and other such problems of patients. Suddenly they would have acquired a universal problem-solving function.

P: There is something in what you say. But we must try to be as fair to the Commission as possible. The report speaks of needs related to quality of life in the event of illness and injury; this is not a question of such needs in general.

I: That does involve a limitation. But the field is still extremely large and the interpretations are many. Are we to mean such needs as are caused by illness or injury? Or are we to mean such as in general terms are attributable to a person with an illness or an injury?

P: I don't think the Commission is interested in making the sharp distinction you indicate. Nor would that be reasonable. We can conceive of health-care staff, in particular hospital social workers, helping patients with problems not directly caused by illness or injury but appearing during the period of care.

I: But then my question is highly relevant. Should health care involve such a large area of responsibility? Another thing: How much is included in quality of life?

P: There you have the decisive question—and the Commission is almost unpardonably vague about it. No analysis of the concept of quality of life is offered. Nor does the Commission indicate a level of quality of life that should be guaranteed. As far as I can see, all that is said on the subject is the following (on p. 98): "Quality of life has physical, mental, social and existential dimensions. Important variables in this connection include, for example, the ability to move about, dress oneself, eat and drink unaided, communicate with others in speech and writing, engage in work/activity and leisure, have viable social relations, satisfactory sleep and the capacity and opportunity for sexual activity."

I: A couple of things immediately strike me when I hear that list of examples. In the first place, the examples appears to refer almost exclusively to an extremely basic quality of life. Which is reasonable. On the list are a number of things which I would think are vital goals (following your terminology) for most people. In the second place, several of the examples concern ability: the ability to move about, to dress oneself, to eat and drink unaided, and to communicate with others. Such abilities come within your concept of health, don't they?

P: Those are correct observations. It is complicated for me to comment on the Commission's proposals inasmuch as we define the basic concepts differently. Several of the things they connect with quality of life, I would connect with health. But anyway we can do a reconstruction in our terms and then see what we think of the proposal.

I: How can we do that?

P: Manifestly, the Commission considers the task of health care not just to attempt to alleviate and cure illness and injury but also to habilitate and re-habilitate people so that they acquire a reasonable degree of health in our sense—in other words, so that they can attain basic vital goals such as being able to move about, to dress themselves, to eat and drink unaided, to communicate with others, to engage in work and leisure pursuits. In cases where this degree of health cannot be independently acquired, the task of health care is to attempt to compensate the patient by helping her or him to attain the vital goals. Such patients should be helped to dress themselves, to communicate with others, to engage in work, etc.

I: Is the level of quality of life that the Commission has in mind close to your minimum happiness? If the patients cannot attain the vital goals on their own, the task of health care is to do it for them.

P: As you can understand, that is a reasonable reconstruction for me. I am not sure that I have thereby interpreted the Commission correctly. But the Commission's examples do speak in favor of my interpretation.

I: So much for quality of life. But what of the Commission's view of health?

P: We have already seen that the Commission appears to have a concept of health which is narrower than mine. But I cannot say anything more precise because the report offers no lucid characterization here either. More clarity occurs when health is set in relation to need for care. On p. 99 of the report, we read the following: "The gravity of a disease can be assessed in various dimensions: the suffering experienced by the patient, the medical prognosis, the functional impairment and the existential distress. The prognosis in turn can refer to the risk of a fatal outcome in the shorter or longer term if care is provided incompletely or not at all, the risk of disability or continuing suffering and impaired quality of life."

I: Are we, then, to understand that the degree of severity of ill-health is to be measured along the dimensions of suffering, disability, and medical prognosis (this last being divided into the dimensions of death, continuing suffering, and disability)? Let me do a diagram. We get the following.

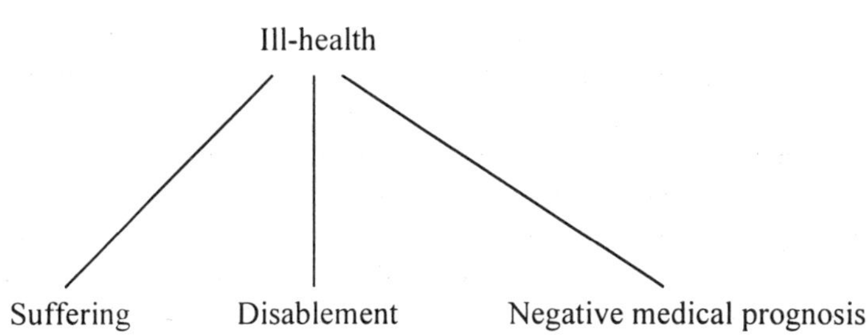

The negative medical prognosis can in turn indicate different things:

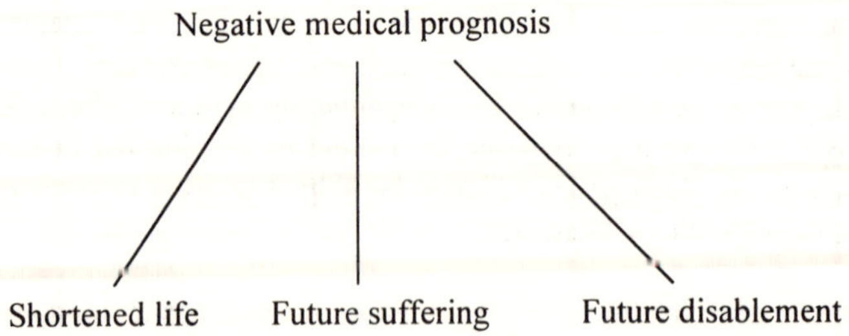

P: Yes, and the complexity becomes enormous when we take into consideration all conceivable lengths of time in the future, because an illness or some type of ill-health can last anything from a minute or two to an entire lifetime.

I: How can we compare all these magnitudes? Does the report help us in any way?

P: It says (p. 99): "It takes a holistic perspective to weigh these various aspects of the gravity of diseases into a yardstick of priorities." The Commission acknowledges the difficulty of such balancing while at the same time emphasizing its necessity.

I: Have you any simple prescription yourself?

P: No, I haven't. I would say, to try to find one would be inadvisable. We risk negative consequences if we introduce mathematized scales and attempt to discover a particular measure of human health and quality of life. Such scales have been in use for evaluating medical technologies. Now comes the idea of using them for assessing need for care.

I: Why do you think the use of such scales involves a risk?

P: Let us consider some of the greatest difficulties. First, we have the scale of suffering, which is forbiddingly difficult to graduate. Even if we limit ourselves to just one sort of suffering, physical pain resulting from an external injury, we see the difficulty in comparing the intensity of one person's pain with that of another person. We have different sorts of pain, too. The pain is different in nature from organ system to organ system and from malady to malady. Furthermore, pain is often accompanied by other suffering, of a more existential nature. This last suffering can in turn vary in nature and strength. Then comes emotional suffering—grief and depression. This, too, can vary in nature and strength. Try then to weigh this diversity of suffering against disability, which is of comparable diversity. In the first place, disability can affect greatly different types of human function. It can affect the locomotive apparatus with all its parts, it can affect the speech organs, and it can affect the mental functions. All these can differ in importance from one person to another, because of differences in people's vital goals. We cannot, therefore, find a uniform measure of the significance of disability. On top of this comes the medical prognosis with its three dimensions—risk of death, risk of continuing suffering, and disability. The same problems—and many additional ones—arise. To crown it all, we have the temporal dimension. You need to be quite a prodigy to discover a figure representing the weighing together of all these disparate factors!

I: But haven't we a number of simplification strategies at our disposal? We need have no illusions about being able to find an exact measure of people's health. The main thing is that we discriminate sufficiently to be able to establish an order of precedence. That is what we need for prioritization.

P: That is a good point. I willingly support simplification strategies so that we can roughly compare the states of health of people. But in that case, we have dropped the ambition I criticize, to produce an absolute measure of a person's health along all the dimensions in question. I am curious as to what simplification strategies you would suggest.

I: A fine starting-point is the person, and the person's own assessment of her or his state of health. This is perfectly in line with your philosophy to say that the person knows best what her or his state of health is. People can themselves determine, at least in qualitative terms, what their suffering is like and to what extent their functions and ability to act have declined.

P: We need expert guidance in the form of the medical prognosis, too. Obviously, we cannot leave the whole of the assessment to the people themselves.

I: Yes—nor do I think that people are fully reliable judges of their degree of disability. People can misjudge both physical and mental functions, so sometimes an external observer needs to step in here as well.

P: So where are we? Let me outline an assessment situation. A young woman goes to a primary care clinic because she has a pain in the joints which makes it difficult for her to do her job as a garbage collector. The physician sees from what the young woman says that the pain is rather severe. The pain is causing a loss of vitality. The physician establishes at the same time that several joint functions, principally in the arms and hands, are impaired. In the course of the conversation, where the young woman describes her job and the different types of physical exertion it requires, and where she also says that she is active in sports (she plays basket ball), the physician and the young woman can together establish that her health is severely impaired. Several essential vital goals are unattainable. The woman receives a thorough medical examination, from which emerges with probability that she has rheumatoid arthritis. Even though the prognosis is not 100 percent certain, this much can be said: virtually no immediate risk of death exists, but quite a large risk exists of lingering, possibly life-long, trouble in the form of pain and immobility in the joints, and also a risk of a worsening of the symptoms in the course of time.

Let us now suppose that the same physician has to deal with a case which in several respects is similar to this one. A young man appears, with the same symptoms. He, too, has a physically demanding job and does sports. But this time, the physician finds no grounds for diagnosing rheumatoid arthritis. He cannot find any serious disease as cause of this man's symptoms. All the indications are that he has a virus infection which has produced symptoms like those of rheumatoid arthritis but which will probably subside within a few days.

I: You have made things easy for yourself.

P: What I wish to indicate by putting these two cases next to each other is that many situations arise where we can say with certainty that one person's need for care is greater than another's need despite a superficial identity. We can elucidate this by reasoning in qualitative terms, without the introduction of mathematics, in order to obtain a measure of health. I admit that in the example of the two young persons the situation is unusual in being so crystal-clear, I admit. What simplifies the comparison is that we can judge the two subjective states of health as being much the same. This means that we can concentrate just on the medical prognosis. And here we find an enormous difference. One of the young persons has a lingering and serious disease, while the other has a rather insignificant and passing disease. Therefore, the first young man has a greater need for care than has the second.

I: Aren't we back at square one? Those intuitive judgments have occurred in medicine throughout the ages.

P: Yes, they have. What is new, though, is the pressure to prioritize between patients in the radical sense that some categories of patient will receive no treatment whatsoever. So we must further sharpen our intuition concerning what constitutes a great need for care.

I: And you have high hopes that such a sharpening is possible?

P: Yes. Indicating the dimensions of need for care, as the Commission has done, and scrutinizing the mutual relations of these dimensions, as you and I have started doing, are steps toward the sharpening of intuition. If this is then followed up in the education of health-care staff, we will have come a long way.

Twenty-One

ON HEALTH AGAIN: CONCLUDING REMARKS

I: You have spent a lot of time answering all my questions and objections concerning your theories of health, quality of life, handicap, and other concepts. Nevertheless, many questions remain. I realize during conversations like these that we could go more deeply on almost every point, finding more problems but also ideas for solving them. All this has been an interesting discovery. I suggest that we devote this last conversation to picking up some major threads from the other conversations—things I regard as in urgent need of further consideration.

P: What do you have in mind?

I: I want to go back to the concept of health. I am keen to do so after our discussion last time on the subject of need for care and the many dimensions of this need. I was struck then that the concept of need for care is more complex than your concept of health. You focus entirely on the dimension of disablement, the inability to act. Doesn't that indicate a limitation in your way of seeing things?

P: Perhaps. But let us have a close look at it. In the first place, need for care is connected with two concepts: health and quality of life. Though we were agreed that the concept of quality of life adopted by the Commission was a limited one, it still goes slightly further than my concept of health. In the second place, remember our discussion of the dimension of suffering (Conversation Fifteen). I do not wish to deny in the slightest the great importance of this component of ill-health. My point was that suffering is *not* a component of *all* ill-health. However, some form of disablement—physical, mental, or social—is to my mind a common feature of ill-health. Sometimes this disablement is a direct consequence of suffering; therefore, suffering comes indirectly into my characterization too.

I: But what of the medical prognosis? The whole temporal horizon is missing from your analysis.

P: I would not agree that the temporal dimension is missing from my analysis. Though I have not included it in my definition of health and ill-health, I have no problem at all in adding it. I can perfectly well speak about ill-health of long duration or of short. I see also as a major and extremely interesting research task to elucidate the different senses of the expression "serious ill-health." Some of these senses have to do with the temporal dimension. Others are directly linked to the degree of disablement, the dimension about which we have talked most.

I: Can you not contemplate introducing a dimension of suffering as a separate dimension in such research? Won't you agree that suffering has a quality which goes far beyond its role as cause of disablement?

P: Oh yes, I'll agree. The degree of suffering is in itself a dimension which we have to take into consideration in assessing the degree of ill-health.

I: But let us consider the following case on which I was prompted to reflect in connection with the prioritization report. A hale and hearty young woman with no aches or pains discovers a small callus in her breast. She goes to see a physician, who diagnoses her as having breast cancer. She undergoes an operation without delay, but it proves impossible to stop the spread of the disease. Her condition rapidly deteriorates, she is in great pain, and soon she dies. Now, how are we to describe this case in your terms? The most problematic thing for me is to characterize her at the beginning. Since she is then hale and hearty, she could be said to be completely healthy in your sense. At the same time, she has a disease with an extremely bad prognosis, indeed the disease proved fatal. Her need for care is the maximum. Do you think all this fits together?

P: I understand your question, but I thought we had sorted that out before. I have said that health is compatible with disease. Health is even compatible, as in your example, with a fatal disease. I have defended this conceptual consequence by saying that it enables us to express key facts. Notice what you yourself said. You used the expression "hale and hearty." You needed an expression to indicate that the young woman felt fine and was perfectly capable of coping with everyday life. But that a person is in good health does not mean that all is well with the person. As in your example, the medical prognosis can be extremely bad. The person can have a disease that calls for immediate treatment. You might say: the woman may be hale and hearty, but all the same she has a great need of care.

I: Why will you not say the woman is hale and hearty? Why do you insist on using the word "health," with its age-old connection with the concept of disease?

P: Now you disappoint me. Has all our discussion been in vain? You seem to think that I have quite arbitrarily chosen the word "health" to designate something which by rights we should call something else. But I have argued at length concerning this—and you have followed and agreed with my argument. There you have the grounds for my position.

I: Sorry if I have lost the hang of things. Could you briefly summarize your argument?

P: I'll be glad to. Let me try to do so as pedagogically as possible. I shall take it one point at a time, skipping all the digressions this time.

1. The starting-point of our whole discussion was the prevalent idea that health is the same as the absence of disease.

2. This idea presupposes that we can define disease without reference to health. Otherwise the definition is circular.

3. We tried defining disease in terms of abnormal physical function or in terms of imbalance. We found no statistical interpretation of normality which could capture the intuitive import of disease. We found no usable interpretation of the concept of equilibrium which did not already presuppose a concept of disease or a concept of health.

4. We chose instead to try to characterize the concept of health without reference to disease. We started from the classical concepts of well-being and ability to act, as also their opposite poles, suffering and disability.

5. We found the category of ability to act to be the most generally usable for a fully comprehensive characterization of health and ill-health.

6. An ability to act must be specifiable in terms of what the person should be able to do. With regard to the concept of health, I chose to define the ability to act in terms of vital goals. I gave a detailed account of the reasoning behind this choice, and you subjected the choice to critical scrutiny.

So a long chain of reasoning, with the concept of health in focus all the time, has brought us to the conclusion that we should link the concept of health with the concept of ability to act.

I: Thank you for that summary. It has been a major step forward for me to see that the concept of need for care is to some extent distinct from the concept of health. A person who is completely healthy can have a great health-related need for care, a need for care which concerns her or his future health.

P: Yes, that's a central point. The concept of need for care is indispensable as a complement to the concepts of health, disease, and quality of life.

I: Here we have found a natural ending for all our discussion. I did get you to tackle a question of current political interest, the question of prioritization in health care. Talking to you has been interesting. You have convinced me on some decisive points, but with regard to other points I am still doubtful and confused.

P: I would have been surpirised, even alarmed, if you had not continued to feel confused. Philosophical reflection never reaches a final goal, and for that reason it must never cease. Fresh angles of approach always await us, and improvement is always possible when we attempt to answer all the large and difficult questions.

I: That sounds awfully pessimistic. Do we never get anywhere? If so, why do we bother to spend time reflecting?

P: You are conflating two things. The fact that we never reach an absolute final point in philosophical reflection (and the same is true regarding empirical science) does not mean that we do not make progress. You and I are much wiser today than we were before all our conversations. But we can be much wiser still, if we carry our efforts beyond this point.

Let me finish, for the present, by thanking you. You have made me think more about things that I had previously neglected. You have made me both waver and feel stronger concerning my ideas. Learning from you—with you—has been useful.

SUMMARY OF CONCEPTS AND THEIR DEFINITIONS

Below I offer more precise definitions of some of the concepts which have come up for consideration in the conversations.

Disability, in the ICIDH definition: In the context of health experience, a disability is any restriction or lack (resulting from an impairment) of ability to perform an activity in the manner or within the range considered normal for a human being.

Disease, in the biostatistical definition: A disease is a state in a person's body or mind which causes subnormal physical function or subnormal mental function.

Disease, in the holistic definition: A disease is a process in a person's body or mind which tends to cause a decline in the person's health (defined holistically—see below). Injury and defect can be given parallel definitions, the one change being the substitution of "state" for "process."

Handicap, in the ICIDH definition: In the health context a handicap is a disadvantage for a given individual resulting from an impairment or a disability that limits or prevents the fulfilment of a role that is normal (depending on age, sex, and social and cultural factors) for that individual.

Happiness, in Nordenfelt's definition: A person P is completely happy if, and only if,

(1) P wishes that (x^1 ... x^n) shall be the case,
(2) (x^1 ... x^n) constitutes the totality of P's wishes, and
(3) P finds that (x^1 ... x^n) is the case.

Happiness, in Tatarkiewiçz's definition: Happiness is a sustained, complete, and justified satisfaction with life as a whole.

Health, in the biostatistical definition: Complete health is the total absence of disease (defined biostatistically—see above).

Health, in Galen's definition: Health is a state in which we suffer no pain and are not prevented from performing any of the tasks of daily life.

Health, in the holistic definition: A person P is completely healthy if, and only if, P has a second-order ability to achieve all her or his vital goals given reasonable circumstances.

Icon of disease, in Sundström's characterization: The iconicity of the integral clinical conceptions (icons) of disease concerns their content or internal make-up, their status as representations and mediators, and their status in the scheme of things.

Illness, in Boorse's definition: A disease is an illness, only if it is serious enough to be incapacitating and therefore is (1) undesirable to its bearer, (2) a title to special treatment, and (3) a valid excuse for normally criticizable behavior.

Illness, in Marinker's definition: Illness is a feeling, an experience of unhealth which is entirely personal, interior to the person of the patient. Sometimes illness exists where no disease can be found.

Illness, in Twaddle's definition: An illness is a subjectively interpreted undesirable state of health. It consists of subjective feeling states (e.g. pain, weakness), perceptions of the adequacy of their bodily functioning, and/or feelings of competence.

Impairment, in the ICIDH definition: In the context of health experience, an impairment is any loss or abnormality of psychological, physiological, or anatomical structure or function.

Need, in Liss's definition: P has a need for x if, and only if,

(1) a difference exists between the present state of affairs and one desirable for P, and
(2) x is a necessary means for elimination of this difference.

Need for care, in Liss's definition: P has a need for care if, and only if,

(1) a difference exists between the present state of affairs and one desirable for P, and
(2) care of P is a necessary means for elimination of this difference.

Need for care, in the Priorities Commission's definition: P has a correspondingly greater need for care the more serious P's illness or injury, or the poorer P's consequent quality of life. P is not in need of measures which do not improve health or quality of life.

Sickness, in Marinker's definition: Sickness is the external and public mode of unhealth. Sickness is a social role, a status, a negotiated position in the world.

Sickness, in Twaddle's definition: Sickness is a social identity. It is the poor health or the health problem(s) of an individual as defined by others with reference to the social activity of that individual.

Vital goal, in Nordenfelt's definition: A vital goal for P is a state the realization of which is a necessary condition for P's minimum happiness.

ANNOTATED BIBLIOGRAPHY

One

On Philosophy of Medicine in general. Solid American introductions to modern philosophy of medicine are the following: H.Tristram Engelhardt and Edmund Erde, "Philosophy of Medicine," in P. T. Durbin (ed.), *A Guide to the Culture of Science, Technology, and Medicine* (New York: Free Press, 1981); and Edmund Pellegrino and David Thomasma, *A Philosophical Basis of Medical Practice* (Oxford: Oxford University Press, 1981). A French philosopher and physician who has played a major role in the development of the philosophy of health (in the francophone world and in the anglophone world) is Georges Canguilhem, through his *Essai sur quelques problèmes concernant le normal et le pathologique* (Clermont-Ferrand: La Montagne, 1943), in English as *On the Normal and the Pathological* (Dordrecht: Reidel, 1978).

Highly readable British introductions to the philosophy of health are David Seedhouse, *Health: The Foundations of Achievement* (Chichester: John Wiley, 1986) and K.W.M. Fulford, *Moral Theory and Medical Practice* (Cambridge, England: Cambridge University Press, 1989).

On Plato (427-347 BC) and Aristotle (384-322 BC). The works of these classical giants of philosophy include contributions to the philosophy of medicine and health. Plato's *Republic*, especially the fourth book, contains reflections on health and health care in general. A good annotated edition is *The Republic* (New York: Loeb Classical Library, 1982 & 1987). Aristotle's comments on health are to be found in, for instance, the *Physics*. Here, as in the *Metaphysics* and the *Nicomachean Ethics*, Aristotle has elaborated his theory of praxis and other concepts of action. All three works are available in the Loeb Classical Library, the respective dates of publication in this series being 1934, 1980, and 1982.

On Galen (ca. 130 - ca. 200). Galen was born in Pergamon in Asia Minor, received his education in Alexandria, and afterwards went to Rome, where he was a physician and an exceptionally prolific author. Galen had a supreme command of contemporary medicine and made an immense contribution to its development. He was also a theorist and philosopher. He was eager to provide his medical thinking with a philosophical basis. Only parts of his writings have been translated into modern languages (chiefly Italian, German, and English). The notion of the latitudes of health is elaborated in his *De temperamentis* (Leipzig: B.G. Teubner, 1904) and *Ars medica. Ars medica* is perhaps the work of his which has been most widely read throughout the ages. It has been translated into German by R. Peters under the title *Über die Heilkunst* (Hamburg: Hamburg University Diss., 1978). A splendid introduction to Galenic medicine is Owsei Temkin, *Galenism: Rise and Decline of a Medical Philosophy* (Ithaca, N.Y.: Cornell University Press, 1973). The dominant role of Galenism in medieval medicine is well described by the Swedish scholar P-G. Ottosson, *Scholastic Medicine and Philosophy* (Naples: Bibliopolis, 1984).

On René Descartes (1596-1650). Descartes was a key figure in the foundation of modern philosophy. His major writings concern epistemology and theory of science. His notion of epistemology was rationalistic in nature: he thought that an extremely large part of our knowledge can be derived from basic conceptions arrived at intuitively. Descartes is well-known for his dualistic ontology. He held that reality is built up of two fundamentally different substances, a bodily and a mental. In his philosophy of health, the main focus is on physical health, which he describes in mechanistic terms. He thereby sharply differentiates health from total well-being, which to a large extent has to do with the mental substance. In his later philosophy, Descartes paid greater attention to the interplay of body and mind. For a scholarly analysis of Descartes's philosophy of medicine, see H. Dreyfus-LeFoyer, "Les conceptions médicales de Descartes," *Revue de Métaphysique et Morale,* 44 (1937), pp. 237-286.

On John Locke (1632-1704) and Thomas Sydenham (1624-1689). Locke was the founder of modern empiricism. His basic notion was that the contents of the mind—its ideas—derive wholly from experience. These ideas, he thought, were the source of all our knowledge of the external world. Less well-known is that the threefold source of inspiration for Locke's empiricist philosophy was his long experience as a physician, his medical research, and his contact with one of the greatest pioneers in the history of medicine, Thomas Sydenham. An account of Locke's work as a physician and of its influence on his philosophy is to be found in Kenneth Dewhurst, *John Locke (1632-1704), Physician and Philosopher: A Medical Biography* (London: Wellcome Historical Medical Library, 1963). Dewhurst has also written a highly readable biography of Sydenham, *Dr. Thomas Sydenham: His Life and Original Writings* (London: Wellcome Historical Medical Library, 1966).

Thomas Sydenham was an extreme empiricist. He detested metaphysical medical systems. He held that all knowledge of diseases should be based on case studies (always of true cases). This has caused him to become a symbol of so-called "bedside medicine." Sydenham wished to avoid the use of general concepts of disease. For Sydenham, no abstract diseases existed which could be designated by general terms. Only individual cases of disease existed. His main work is *Observationes medicae* (London: G. Kettilby, 1676).

On Friedrich Nietzsche (1844-1900). Nietzsche became Professor of Classical Philology in Basel while he was still a young man. However, he soon broke off his traditional academic career and produced a number of pioneering yet contradictory writings on science, religion, and morality. Well-known are his polemic against Christianity and his appeal for a heroic morality, this going with a contempt for weakness. In some of his books, most notably *Die Fröhliche Wissenschaft* (Munich: Goldmann, 1959), he elaborates ideas which anticipate a holistic theory of health. On pages 176-177 of Walter Kaufmann's translation, *The Gay Science* (New York: Random House, 1974), we find, under the rubric "Health of the soul," the following:

> For there is no health as such, and all attempts to define a thing that way have been wretched failures. Even the determination of what is healthy for your *body* depends on your goal, your horizon, your energies, your impulses, your errors, and above all on the ideals and phantasms of your soul. Thus there are innumerable healths of the body; and the more we allow the unique and incomparable to raise its head again, and the more we abjure the dogma of the "equality of men," the more must the concept of a *normal* health, along with a normal diet and the normal course of an illness, be abandoned by medical men.

On Michel Foucault. Foucault made his name as a philosopher with *Folie et déraison: Histoire de la folie à l'âge classique* (Paris: Plon, 1961), in English as *Madness and Civilization: A History of Insanity in the Age of Reason* (New York: Pantheon Books, 1965). A later though opaque work is *Naissance de la clinique: Une archéologie du regard médical* (Paris: Presses universitaires de France, 1963), in English as *The Birth of the Clinic: An Archeology of Medical Perception* (New York: Pantheon Books, 1973).

On medical ethics and the role of hospital ethicist. Not much has yet been written on medical ethicists in their role as practical hospital ethicists. See, though, David Blake, "The Hospital Ethics Committee: Health Care's Moral

Conscience or White Elephant?", *Hastings Center Report* (January-February 1992).

On modern medical history. One of the best accounts of the medical revolution during the late years of the nineteenth century and the early years of the twentieth is Richard H. Shryrocke, *The Development of Modern Medicine: An Interpretation of the Social and Scientific Factors Involved* (Madison: University of Wisconsin Press, 1979).

Two

On interpretation and hermeneutics. Richard Palmers, *Hermeneutics* (Evanston, Ill.: Northwestern University Press, 1969) is a highly readable and thorough introduction to such classical figures of hermeneutics as Wilhelm Dilthey (1833-1911), Martin Heidegger (1899-1976), and Hans-Georg Gadamer (1902-). A modern presentation in German is Helmut Seifert, *Einführing in die Hermeneutik* (Tübingen: Francke, 1992).

On diagnostic criteria and medical concepts. The US/UK diagnostic project is carefully described in R. E. Kendell, *The Role of Diagnosis in Psychiatry* (Oxford: Blackwell, 1975). The final report on the project itself—a project with considerably more elements than I mention—is: J. E. Cooper *et al.*, "Psychiatric Diagnosis in New York and London" (London: Maudsley Monograph No. 20, 1972). Kendell's book provides many more illustrations of the problem of uniformity with regard to medical concepts. Conceptual problems in connection with statements of cause of death and with cause of death statistics are dealt with in my *Causes of Death: A Philosophical Essay* (Stockholm: Swedish Council for Planning and Coordination of Research, 1983).

On the international standardization of concepts. Providing a standard for the nomenclature and classification of diseases is the *International Statistical Classification of Diseases and Related Health Problems*, normally abbreviated as ICD. A new edition of this manual has appeared about every ten years. The latest is the tenth edition (ICD-10), published by WHO in Geneva in 1992. It covers all types of health problem. It is supplemented in the sphere of psychiatry in many countries by the *Diagnostic and Statistical Manual of Mental Disorders*, published by the American Psychiatric Association. This manual is now in its fourth edition (*DSM-IV*, Washington: The American Psychiatric Association, 1994).

On Immanuel Kant (1724-1804). Kant was professor in Königsberg, East Prussia, where he lived all his life. His theoretical philosophy revolutionized Western philosophical thought. The Swedish philosopher Konrad Marc-Wogau has summed up Kant's basic epistemological idea as "it is not reason that complies with nature but nature that complies with reason." Kant elaborated this idea in *Kritik der Reinen Vernunft* (Riga: J.F. Hartknoch, 1781), in English, *The Critique of Pure Reason*, translated by J.M.D. Meiklejohn (Chicago: Encyclopaedia Britannica, 1955), and the more concise *Prolegomena zu einer jeden künftigen metaphysik die als wissenschaft wird auftreten können* (Riga: J.F. Hartknoch, 1783).

Three

On conceptual analysis. Philosophers of the so-called Oslo school, with Arne Naess at their head, have a view of conceptual analysis and the study of concepts which differs from the one I have presented. Naess strongly advocates empirical studies of linguistic usage for the purpose of establishing what terms mean. Naess's main work is *Interpretation and Preciseness: A Contribution to the Theory of Communication* (Oslo: Gröndahl & Son, 1953).

On Wittgenstein and the family concepts. Ludwig Wittgenstein (1889-1951) is among the twentieth-century philosophers about whom most has been written. His work has caught the interest of representatives of the most diverse ways of philosophizing. His thought underwent a singular development, and during the later part of his life he repudiated the philosophy of his youth. Wittgenstein's theory of the family concepts, which belongs to his later period, is found in his *Philosophical Investigations* (Oxford: Blackwell, 1953). An enormous amount of literature exists on Wittgenstein. Occupying a place apart is G. P. Baker & P. M. S. Hacker, *An Analytical Commentary on the Philosophical Investigations*, the fifth and last part of which is being published by Oxford University Press. A highly readable first-rate introduction is Anthony Kenny, *Wittgenstein* (London: Lane, 1973).

On Austin and linguistic philosophy. John Langshaw Austin (1911-1960) was the leader during the 1950s of the so-called Oxford school of philosophy. Central to this school's approach to philosophy was a concentration on the everyday use of philosophically relevant terms. It is often referred to as ordinary language philosophy. Austin frequently began by consulting an advanced dictionary. His best-known book is *How to Do Things with Words,* ed. by J.O. Urmson and Marina Sbisa (Cambridge, Mass.: Harvard University Press, 1975).

On Plato's philosophical method. The Theaetetus dialogue. Almost all the extant writings of Plato are in dialogue form. In several of them, Socrates is the protagonist. Socrates' technique is always to act ignorant regarding the problem to be discussed, letting the other person repeatedly display an apparently impregnable certainty of opinion—which Socrates then elegantly breaches. The *Theaetetus* dialogue is available in Penguin Classics: *Theaetetus*, translated with an essay by Robin A. H. Waterfield (Harmondsworth, 1987).

On sociological studies of concepts. A modern sociological analysis of notions of health is Gunilla Tegern, *Frisk och sjuk: Vardagliga föreställningar om hälsan och dess motsatser* (Healthy and Ill: Everyday Notions of Health and Its Opposites) (Linköping: Linköping Studies in Arts and Science, No.119, 1994). Published in Swedish with an English summary. Tegern's approach is based in part on the sociological theorizing concerning so-called social representations. By a social representation is meant a mental image—often a simplified one—of a phenomenon which is shared and kept alive within a cultural community. Prominent figures in this theorizing are S. Moscovici and R. M. Farr; see their *Social Representations* (Cambridge, England: Cambridge University Press, 1984). The Mugwe figure mentioned in the third conversation has been the subject of an exhaustive anthropological study by Bernardo Bernardi, *The Mugwe, a Failing Prophet: A Study of a Religious and Public Dignitary of the Meru of Kenya* (Oxford: Oxford University Press, 1959).

Four

On the International Classification of Impairments, Disabilities, and Handicaps (ICIDH). This is a manual whose purpose is like that of *ICD*, namely, to provide health-care staff with an internationally valid nomenclature and classification. It was published by the WHO in Geneva, 1980. (See Conversation Eighteen.)

On definitions in medical science. Karl R. Popper (1902-1994) strongly criticizes the idea that all—or even the vast majority of—scientific concepts need to be defined; see, *e.g.*, his *Conjectures and Refutations: The Growth of Scientific Knowledge* (London: Routledge & Kegan Paul, 1963). The idea that all the concepts should be defined is absurd, since certain of them must remain undefined for logical reasons. A chain of definition must always have an undefined concept from which to start. But Popper goes further, regarding as unfruitful to seek to reduce the number of undefined concepts to a minimum: "We need many undefined terms whose meaning is only precariously fixed by usage—by the manner in which they are used in the context of theories, and by the procedures and practices of the laboratory" (*Conjectures and Refutations*, p. 279). Uffe Juul Jensen is a Danish philosopher who has also interestingly called into question the need for definitions in science; see his *Practice and Progress: A Theory for the Modern Health-Care System* (Oxford: Blackwell, 1987).

Five

On the WHO definition of the concept of health. This definition comes right at the start of WHO's constitution, which was adopted on 22 July 1946. It reads as follows:

> Health is a state of complete physical, mental and social well-being and not merely the absence of disease and infirmity. (*Official Records of the World Health Organization*, 2 (1948), p. 100 [Geneva])

On biostatistical concepts of health and disease. Among the most lucid attempts to define concepts of health and disease in modern times are the essays by the American philosopher Christopher Boorse which appeared in the 1970s. The most central of them is "Health as a Theoretical Concept," *Philosophy of Science,* 44 (1977), pp. 542-573. Much criticism of Boorse has occurred, a comprehensive example of which is found in K.W.M. Fulford, *Moral Theory and Medical Practice* (Cambridge, England: Cambridge University Press, 1989).

Six

On normal values and reference values. Clinical chemists nowadays prefer to speak of "reference values" rather than "normal values," partly for reasons presented in this conversation. For a solid introduction to the subject, see Ralph Gräsbeck and Torgny Alström, *Reference Values in Laboratory Medicine: The Current State of the Art* (Chichester: John Wiley, 1981). See also Otto Klineberg's classic study of cultural variations in respect of somatic reference values, *Tensions Affecting International Understanding: A Survey of Research* (New York: Social Science Research Council, Bulletin 62, 1950). One of the most profound studies of the concept of normality is Georges Canguilhem, *On the Normal and the Pathological* (Dordrecht: Reidel, 1978). A survey of concepts of normality which can be of relevance to medicine is found in Edmond Murphy, *The Logic of Medicine* (Baltimore: Johns Hopkins University Press, 1976), ch. 9.

Seven

On equilibrium and homoeostasis. The classical doctrine of equilibrium as a theory of health and disease is of ancient origin, appearing in the writings of Hippocrates, the founder of medical ethics, who lived from 466 to 377 BC. It is especially well developed in the writings of Galen. According to Galen, equilibrium, that is health, is present if the primary qualities of the body—heat, cold, dryness, and wetness—balance one another. In concrete terms, the doctrine was interpreted as stating that the body's fluids should be in the right proportions to one another. These fluids were taken to be four in number: blood, phlegm, yellow bile, and black bile. For Galen and other theorists of antiquity, the blood was hot and wet, the phlegm was cold and wet, the yellow bile was hot and cold, and the black bile was cold and dry. Walter Cannon's modern theory of equilibrium, interpreted in terms of the body's homoeostatic functions, is found in his *The Wisdom of the Body* (New York: Norton, 1932).

On the HUGO project and genetic mapping. The gigantic work now being carried out to map the human gene pool is well described in Walter Bodmer and Robin McKie, *The Book of Man* (London: Little, Brown & Co, 1994). For a broader description, which also considers ethical and economic aspects, see Joel Davis, *Mapping the Code: The Human Genome Project and the Choices of Modern Science* (Chichester: John Wiley, 1990).

Eight

On icons of disease. The ingenious use of the term "icon" to designate a clinical conception of disease was introduced by Per Sundström in his doctoral dissertation *Icons of Disease: A Philosophical Inquiry into the Semantics, Phenomenology, and Ontology of the Clinical Conceptions of Disease* (Linköping: Linköping Studies in Arts and Science, No.14, 1987).

On insanity. An introduction to the history of insanity is the above-mentioned *Madness and Civilization* by Michel Foucault (English edition, London: Tavistock Publications, 1967). Both polemical and influential is Thomas Szasz, *Ideology and Insanity: Essays on the Psychiatric Dehumanization of Man* (London: Anchor Books, 1970). An analytically penetrating study of insanity from both a philosophical and a sociological perspective is Jeff Coulter, *Approaches to Insanity* (London: Robertson, 1973).

On schizophrenia. This is the best-known and probably the most discussed of all forms of psychosis. In *DSM-IV* the most pregnant characterization of schizophrenia is the following:

> Schizophrenia is a disturbance that lasts for at least 6 months and includes at least one month of active-phase symptoms (*i.e.* two or more of the following: delusions, hallucinations, disorganized speech, grossly disorganized or catatonic behaviour, negative symptoms). (Schizophrenia can be subdivided into Paranoid, Disorganized, Catatonic, Undifferentiated and Residual.) (p. 273)

On manic-depressive psychosis. This is the traditional designation of the form of psychosis involving an alternation betwen a manic, hyperactive state and a depressive state. In *DSM-IV* a manic state is described as follows:

> Manic episode is defined by a distinct period during which there is an abnormally and persistently elevated, expansive, or irritable mood. This period of abnormal mood must last at least one week. The mood disturbance must be accompanied by at least three additional symptoms from a list which includes inflated selfesteem or grandiosity, decreased need for sleep, pressure of speech, flight of ideas, distractibility, increased involvement in goal-directed activities or psychomotor agitation, and excessive involvement in pleasurable activities with a high potential for painful consequences. (p. 328)

In *DSM-IV,* the term "manic-depressive psychosis" has been replaced by "bipolar mood disorder."

Nine

On Plato and Cicero and mental health. The idea that virtue and mental health are one and the same thing is a pervasive feature of the world-view of antiquity. We find the idea, for instance, in the fourth book of Plato's *Republic*. It was perhaps best elaborated by the Stoics. The most famous Stoic, Chrysippos (286-204 BC), based what amounted to a whole system of psychiatry on the idea. For Chrysippos and the Stoics in general, the distinctive mark of virtue was living in a passionless (apathetic) state. They looked upon strong emotions as disturbances of the psyche. Marcus Tullius Cicero (106-43 BC) gives a vivid description of the doctrine in his *Tusculan Disputations*

(New York: Loeb Classical Library, 1934). We find a fine survey of the thinking of late antiquity on this in Martha Nussbaum, *The Therapy of Desire: Theory and Practice in Hellenistic Ethics* (Princeton: Princeton University Press, 1994).

On psychiatry in the USSR. A description and analysis of the state of affairs in Soviet psychiatry is provided by S. Bloch in his essay, "The Political Misuse of Psychiatry in the Soviet Union," in S. Bloch and P. Chodoff, *Psychiatric Ethics* (Oxford: Oxford University Press, 1981). See also K.W.M. Fulford's essay written in collaboration with A. Y. U. Smirnov and E. Snow, "Concepts of Disease and the Abuse of Psychiatry in the USSR," *British Journal of Psychiatry*, 162 (1993), pp. 801-810. In 1977 in Hawaii the Sixth World Psychiatric Assembly adopted an international code of ethics for psychiatric care and therapy. Here we have a crystal-clear rejection of the political misuse of psychiatry.

On the measurement of mental health. Stephen Platt, "Social Adjustment as a Criterion of Treatment Success: Just What Are We Measuring?", *Psychiatry*, 44 (1981), pp. 95-112, has drawn attention to the question of what we in truth are measuring when we think we are measuring mental health.

On psychiatry and homosexuality. Ron Bayer, *Homosexuality and American Psychiatry: The Politics of Diagnosis* (New York: Princeton University Press, 1981), offers a vivid and detailed account of the prehistory of the American Psychiatric Association's decision on the status of homosexuality.

Ten

On Galen and the health dimension. Per Gunnar Ottosson, *Scholastic Medicine and Philosophy* (Naples: Bibliopolis, 1984) provides a survey of Galen's doctrine of the latitudes of health. An extremely readable introduction to Galen's medicine is O. Temkin, *Galenism: Rise and Decline of Medical Philosophy* (Ithaca, N.Y.: Cornell University Press, 1973).

On the relationship between diseases and health. A line of reasoning attempting to indicate the relationship between diseases and health is found in my *On the Nature of Health* (especially ch. 4). Similar views are to be found in Ingmar Pörn, "An Equilibrium Model of Health" in Lennart Nordenfelt and B. Ingemar B. Lindahl (eds.), *Health, Disease, and Causal Explanations in Medicine* (Dordrecht: Reidel, 1984).

Eleven

On philosophical psychology. Several questions which arise in this conversation belong to the field of philosophical psychology. This is a branch of philosophy which has acquired an extremely strong position since the end of World War II. Two works of importance are Gilbert Ryle, *The Concept of Mind* (London: Hutchinson, 1949), and Ludwig Wittgenstein, *Philosophical Investigations* (published posthumously, Oxford: Blackwell, 1953). In addition to consideration of the basic question of the nature of mind and its relation to the body, this tradition involves consideration of how the different mental categories are to be characterized and distinguished from one another. Important in this respect is Anthony Kenny, *Action, Emotion, and Will* (London: Routledge & Kegan Paul, 1963). My reflections are based on Kenny's book and on some later works, for instance, O. H. Green, *The Emotions: A Philosophical Theory* (Dordrecht: Kluwer, 1992). For a more thorough exposition concerning sensations and emotions as discussed here, see my *On the Nature of Health,* pp.82-86.

On the holistic theory of health. The original definition of my holistic theory of health, together with the reasons behind it, is found in *On the Nature of Health*, pp. 46-96. In the Appendix to the second edition (1995), I offer a defense of this theory in relation to three theories which are similar to it. The definition of health presented in the eleventh conversation, with its reference to "reasonable" circumstances, is a slightly revised version of the original one. The reasoning behind the revision is to be found in ch. 9 of my *Action, Ability, and Health* (forthcoming).

On the concept of need. This has been thoroughly studied in the modern literature. A *locus classicus* is K.E.Tranöy, "Ought Implies Can: A Bridge from Fact to Norm," Part 2, *Ratio*, 17 (1975), pp. 147-175. A good later analysis is to be found in Per-Erik Liss, *Health Care Need* (Aldershot: Avebury, 1993). A solid work which takes up both conceptual-analytical and ethical questions is David Braybrooke, *Meeting Needs: Studies in Moral, Political, and Legal Philosophy* (Princeton: Princeton University Press, 1987).

Twelve

On decisions, intentions, and goals. For an analysis of different sorts of goal in the sphere of health, see my *On the Nature of Health,* pp. 70-80. Concerning the distinction between decisions, intentions, wishes, and other concepts of will, see my *Explanation of Human Actions* (Uppsala: Department of Philosophy, Uppsala University, 1974).

Thirteen and Fourteen

On happiness and quality of life. The history of philosophy contains a host of writings on happiness. Two of the best known from antiquity are Aristotle, *Nicomachean Ethics* (London: Loeb Classical Library, 1934), and Cicero in the fifth of his *Tusculan Conversations*. Two influential modern studies are Władysław Tatarkiewicz, *Analysis of Happiness* (The Hague: Nijhoff, 1976), and Elizabeth Telfer, *Happiness* (London: Macmillan, 1980). For a classic account of utilitarianism and its concept of happiness, see Jeremy Bentham, *An Introduction to the Principles of Morals and Legislation,* ed. J. H. Burns and H. L. A. Hart (London: Methuen, 1970/1982). A broad general survey from the perspective of the history of ideas is V. J. McGill, *The Idea of Happiness* (New York: Praeger, 1967). One of the most comprehensive works on the subject in the tradition of empirical psychology is Ruth Veenhoven, *Conditions of Happiness* (Dordrecht: Reidel, 1984). The modern literature on quality of life in general and health-related quality of life in particular is enormous. For an introduction, see the bibliographies in my *Concepts and Measurement of Quality of Life in Health Care* (Dordrecht: Kluwer, 1994). Some of the complications which lie concealed in the concept of happiness and which are discussed in this conversation receive more thorough treatment in my *Quality of Life, Health, and Happiness* (Aldershot: Avebury, 1993).

Fifteen

On the trichotomy disease, illness, and sickness. Many writers in the medical, the philosophical, and the social-scientific fields make a basic distinction between disease and illness, though not all in the same way. In *Quality of Life, Health, and Happiness,* pp. 103-112, I give an account of four different grounds for the distinction. A renowned medical work where the concept of illness plays a major role is A.R. Feinstein, *Clinical Judgment* (New York: Krieger, 1967). The concept of sickness has acquired a central position in modern sociological and anthropological literature. In brief, sickness can be said to be a question of the role people assume when they recognize themselves as being ill and are also looked upon as ill by those around them. A classic essay on the subject is A.Young, "The Anthropologies of Illness and Sickness," *Annual Review of Anthropology,* 11 (1982), pp. 257-285. A critical discussion of all three concepts is Andrew Twaddle and Lennart Nordenfelt, *Disease, Illness, and Sickness: Three Central Concepts in the Theory of Health* (Linköping: Linköping Studies on Health and Society, No. 18, 1993).

Sixteen

On modern theories of health. Among European analyses of the concept of health in holistic terms are K.W.M. Fulford, *Moral Theory and Medical Practice* (Cambridge, England: Cambridge University Press, 1989), David Seedhouse, *Health: Foundations of Achievement* (Chichester: John Wiley, 1986), and Ingmar Pörn, "Health and Adaptedness," *Theoretical Medicine,* 14 (1993), pp. 295-303. An American essay on the same lines is Caroline Whitbeck, "A Theory of Health" in A. Caplan and H.T. Engelhardt (eds.) *Concepts of Health and Disease: Interdisciplinary Perspectives* (Reading, Mass.: Addison-Wesley, 1981).

Seventeen

On maladies. The Americans C.M. Culver and E. Gert have introduced the use of "maladies" as a generic term for the entire sphere of pathology, covering diseases, injuries, and defects. Their main work is *Philosophy in Medicine: Conceptual and Ethical Issues in Medicine and Psychiatry* (Oxford: Oxford University Press, 1982). *On universality in medical concepts.* For a more thorough discussion of the possibility of universal concepts of disease against the background of a holistic theory of health, see my *On the Nature of Health,* pp. 129-131.

Eighteen

On environmental care and "goal care." In Swedish, the use of these terms in connection with health care is dealt with in Ingmar Pörn, "Vårdlära och vårdvetenskap: Några vetenskapsteoretiska synpunkter" (Theory and Science of Care: Some Scientific-Theoretical Views), *Socialmedicinsk Tidskrift,* 7 (1984), pp. 302-306.

Nineteen

On the International Classification of Impairments, Disabilities, and Handicaps (ICIDH). The first edition of this classification, published by the WHO in 1980, was of an experimental nature, but since 1993 it has been an official publication. The three concepts—impairment, disability, and handicap—are seen as related to one another like the links in a causal chain. Impairments can cause disabilities, which in turn can cause handicaps. In rough terms, impairments have to do with the anatomical/physiological plane, disabilities have to do with the integrated locomotor plane, and handicaps have to do with the ability of people to perform their roles in society. Several criticisms of the ICIDH have been raised, and improvements have been suggested. Especially active in this respect has been the Canadian Society for the ICIDH. See their "Comments on the Proposition of Revision of the Handicap Concept" in *ICIDH International Network,* 3 (1990). I have critically examined the ICIDH in "On the Notions of Disability and Handicap," *Scandinavian Journal of Social Welfare,* 2 (1993), pp. 17-24.

On the concept of role and the concept of the sick role. The concept of role occupies a central position in modern social psychology. It is closely linked to the theory known as structural functionalism, which asserts that societies and social systems are to be seen as organic units where different parts fulfill different given functions and work together for the equilibrium and harmony of the whole. People to a large extent fulfill their functions through the assumption of different roles. A role is defined—according to structural functionalism—by a series of norms which indicate what the person who has assumed the role must do, is allowed to do, and is not allowed to do. The classic theorist in this field is the American Talcott Parsons. His principal work is *The Social System* (New York: The Free Press, 1964).

Talcott Parsons is also the originator of the idea of a specific sick role. What this idea involves is that when people have acknowledged that they are sick they assume a sick role which is defined by the following norms, two of them rights and two of them duties:

(1) The person can abandon her or his usual social roles.
(2) The person cannot be held responsible for the sickness, cannot be expected to become healthy through an effort of will.
(3) The person should strive to become healthy.
(4) The person should seek competent help and cooperate all the time with the person or persons giving such help.

Parsons originally presented his idea of the sick role in the above-mentioned *The Social System*. Andrew Twaddle has elaborated and refined the concept in many writings, chiefly in *Sickness Behavior and the Sick Role* (Boston: G.K. Hall, 1979).

Twenty

On the Priorities Commission. The report discussed in this conversation is *Vårdens svåra val: Slutbetänkande av Prioriteringsutredningen* (*Priorities in Health Care: Ethics, Economy , and Implementation. Final Report by The Swedish Parliamentary Priorities Commission)* (Stockholm: Swedish Government Official Reports, Ministry of Health and Social Affairs, 1995, No.5). In the Commission's terms of reference the principal task laid down was "to discuss the role of health services in the welfare society and to highlight the basic ethical principles by which necessary prioritisation ought to be guided." The Commission reported its task as having been "to structure the problems and indicate values by which those who have to decide priorities, both at political/administrative level and in everyday clinical practice, should be guided." Heading the commission was Jerzy Einhorn, professor and member of parliament. Its work was completed in 1995, when the final report was presented.

On the measuring of health. An enormous literature exists on the measuring of health, and on the measuring of quality of life. A useful survey of instruments for the measurement of health is offered in I. McDowell and C. Newell, *Measuring Health: A Guide to Rating Scales and Questionnaires* (Oxford: Oxford University Press, 1987).

Summary of Concepts and Their Definitions

C. Boorse's definitions of disease, health, and illness exist in, for instance, "Health as a Theoretical Concept," *Philosophy of Science*, 44 (1977), pp. 542-573.

L. Nordenfelt's definitions of disease, health and other concepts can be found in *On the Nature of Health*, concerning health, pp.76-80, and concerning disease, injury and defect, pp.105-112. J.H. Marinker's definitions of disease, illness and sickness exist in "Why Make People Patients?", *Journal of Medical Ethics*, 1 (1975), pp. 81-84.

A. Twaddle's definitions of disease, illness, and sickness exist in A. Twaddle and L. Nordenfelt, *Disease, Illness, and Sickness: Three Central Concepts in the Theory of Health* (Linköping: Linköping University Studies on Health and Society, No.18, 1993).

Galen's definition of health is reproduced in O. Temkin, "A Scientific Approach to Disease: Specific Entity and Individual Sickness," in A.C. Crombie (ed.), *Scientific Change: Historical Studies in the Intellectual, Social, and Technical Conditions for Scientific Discovery and Technical Invention from Antiquity to the Present* (New York: Basic Books Inc., 1963).

W. Tatarkiewiçz's definition of happiness exists in his *Analysis of Happiness* (The Hague: Martinus Nijhoff, 1976), p. 8.

L. Nordenfelt's definition of happiness exists in his *Quality of Life, Health, and Happiness* (Aldershot: Avebury, 1993), p. 45.

P. Sundström's characterization of icons of disease can be found in his *Icons of Disease: A Philosophical Inquiry into the Semantics, Phenomenology, and Ontology of the Clinical Conceptions of Disease* (Linköping: Linköping Studies in Arts and Sciences, No.14, 1987), p.132.

P.-E. Liss's definitions of need and need for care can be found in his *Health Care Need* (Aldershot: Avebury, 1993), p. 50 and p.85.

The definition of need for care by the Priorities Commission is to be found in their report *Priorities in Health Care: Ethics, Economy, Implementation* (Stockholm: Swedish Government Official Reports, The Ministry of Health and Social Affairs, 1995, No. 5), p. 99.

L. Nordenfelt's definition of vital goal exists in *On the Nature of Health,* pp. 90-92.

The WHO's definitions of impairments, disabilities, and handicaps exist in their document *International Classification of Impairments, Disabilities, and Handicaps* (Geneva: WHO, 1980).

ABOUT THE AUTHOR

Lennart Nordenfelt is Professor of Philosophy of Medicine and Health Care at the Department of Health and Society, University of Linköping, Sweden. He studied philosophy, history of ideas, history of religion, and Russian at Uppsala University. As a graduate student he spent a year, 1969, at Oxford, under the supervision of John Mackie, where he started preparing his doctoral dissertation on explanation of human actions, which he defended at Uppsala in 1974. After a year as a secretary at the Office of the Chancellor of the Swedish Universities, he was appointed in 1975 a lecturer in Theoretical Philosophy at the University of Stockholm. From 1978 onwards, Nordenfelt has devoted himself to studies and research in the philosophy of medicine and health care. The new research interest prompted him to apply for a position as Associate Professor at the then newly founded transdisciplinary faculty of themes at Linköping University in 1982. He initiated there a research program under the title Theory, Ethics, and Ideology of Health Care which is still running. Seven persons have so far taken their PhDs within this program. He became a full professor in 1987. He and his wife Kerstin have two children.

Nordenfelt's monographs include *Explanation of Human Actions* (1974), *Events, Actions, and Ordinary Language* (1977), *Causes of Death* (1983), *On the Nature of Health* (1987; second, revised, edition 1995), *On Crime, Punishment, and Psychiatric Care* (1992), *Quality of Life, Health, and Happiness* (1993), and *Action, Ability, and Health* (forthcoming). Together with B. Ingemar B. Lindahl, he edited *Health, Disease, and Causal Explanations in Medicine* (1984). This book was the academic result of the First Nordic Symposium on the Philosophy of Medicine, held in Stockholm, 1982. He is the editor of *Concepts and Measurements of Quality of Life in Health Care* (1994). Nordenfelt is also the author of a number of monographs and textbooks in Swedish.

He has contributed to several journals, including *Theoria, Theoretical Medicine, Philosophy, Psychiatry & Psychology, Health Care Analysis, International Journal of Technology Assessment in Health Care, and*

Scandinavian Journal of Social Medicine. He was a research fellow at the University of Western Ontario, London, Canada, in 1976, and at the Institute for Advanced Studies in the Humanities at Edinburgh University, Scotland, in 1989. He has made presentations to philosophical as well as medical gatherings in all the Scandinavian countries and in England, Scotland, the Netherlands, France, Austria, Hungary, the United States, and Canada. He is a member of the executive committee of the European Society for Philosophy of Medicine and Health Care. He is a co-editor of the European Studies in Philosophy of Medicine, published by Kluwer Academic Publishers, Dordrecht, the Netherlands. He has since 1994 been a member of the board of Linköping University.

Nordenfelt has made contributions mainly to action theory and the theory of health and welfare. His theory of health draws upon concepts central to action theory. The concept of health itself is in his theory defined in terms of the subject's ability to realize her or his vital goals. This book contains elaborations and defenses of this theory of health.

INDEX

VIBS

The **Value Inquiry Book Series** is co-sponsored by:

American Maritain Association
American Society for Value Inquiry
Association for Personalist Studies
Association for Process Philosophy of Education
Center for East European Dialogue and Development, Rochester Institute of Technology
Centre for Cultural Research, Aarhus University
College of Education and Allied Professions, Bowling Green State University
Concerned Philosophers for Peace
Conference of Philosophical Societies
International Academy of Philosophy of the Principality of Liechtenstein
International Society for Universalism
International Society for Value Inquiry
Natural Law Society
Philosophical Society of Finland
Philosophy Seminar, University of Mainz
R.S. Hartman Institute for Formal and Applied Axiology
Society for Iberian and Latin-American Thought
Society for the Philosophic Study of Genocide and the Holocaust
Yves R. Simon Institute.

Titles Published

1. Noel Balzer, *The Human Being as a Logical Thinker.*

2. Archie J. Bahm, *Axiology: The Science of Values.*

3. H. P. P. (Hennie) Lötter, *Justice for an Unjust Society.*

4. H. G. Callaway, *Context for Meaning and Analysis: A Critical Study in the Philosophy of Language.*

5. Benjamin S. Llamzon, *A Humane Case for Moral Intuition.*

6. James R. Watson, *Between Auschwitz and Tradition: Postmodern Reflections on the Task of Thinking.* A volume in **Holocaust and Genocide Studies.**

7. Robert S. Hartman, *Freedom to Live: The Robert Hartman Story,* edited by Arthur R. Ellis. A volume in **Hartman Institute Axiology Studies.**

8. Archie J. Bahm, *Ethics: The Science of Oughtness.*

9. George David Miller, *An Idiosyncratic Ethics; Or, the Lauramachean Ethics.*

10. Joseph P. DeMarco, *A Coherence Theory in Ethics.*

11. Frank G. Forrest, *Valuemetrics: The Science of Personal and Professional Ethics.* A volume in **Hartman Institute Axiology Studies.**

12. William Gerber, *The Meaning of Life: Insights of the World's Great Thinkers.*

13. Richard T. Hull, Editor, *A Quarter Century of Value Inquiry: Presidential Addresses of the American Society for Value Inquiry.* A volume in **Histories and Addresses of Philosophical Societies.**

14. William Gerber, *Nuggets of Wisdom from Great Jewish Thinkers: From Biblical Times to the Present.*

15. Sidney Axinn, *The Logic of Hope: Extensions of Kant's View of Religion.*

16. Messay Kebede, *Meaning and Development.*

17. Amihud Gilead, *The Platonic Odyssey: A Philosophical-Literary Inquiry into the* Phaedo.

18. Necip Fikri Alican, *Mill's Principle of Utility: A Defense of John Stuart Mill's Notorious Proof.* A volume in **Universal Justice.**

19. Michael H. Mitias, Editor, *Philosophy and Architecture.*

20. Roger T. Simonds, *Rational Individualism: The Perennial Philosophy of Legal Interpretation.* A volume in **Natural Law Studies.**

21. William Pencak, *The Conflict of Law and Justice in the Icelandic Sagas.*

22. Samuel M. Natale and Brian M. Rothschild, Editors, *Values, Work, Education: The Meanings of Work.*

23. N. Georgopoulos and Michael Heim, Editors, *Being Human in the Ultimate: Studies in the Thought of John M. Anderson.*

24. Robert Wesson and Patricia A. Williams, Editors, *Evolution and Human Values.*

25. Wim J. van der Steen, *Facts, Values, and Methodology: A New Approach to Ethics.*

26. Avi Sagi and Daniel Statman, *Religion and Morality.*

27. Albert William Levi, *The High Road of Humanity: The Seven Ethical Ages of Western Man,* edited by Donald Phillip Verene and Molly Black Verene.

28. Samuel M. Natale and Brian M. Rothschild, Editors, *Work Values: Education, Organization, and Religious Concerns.*

29. Laurence F. Bove and Laura Duhan Kaplan, Editors, *From the Eye of the Storm: Regional Conflicts and the Philosophy of Peace.* A volume in **Philosophy of Peace.**

30. Robin Attfield, *Value, Obligation, and Meta-Ethics.*

31. William Gerber, *The Deepest Questions You Can Ask About God: As Answered by the World's Great Thinkers.*

32. Daniel Statman, *Moral Dilemmas.*

33. Rem B. Edwards, Editor, *Formal Axiology and Its Critics.* A volume in **Hartman Institute Axiology Studies.**

34. George David Miller and Conrad P. Pritscher, *On Education and Values: In Praise of Pariahs and Nomads.* A volume in **Philosophy of Education.**

35. Paul S. Penner, *Altruistic Behavior: An Inquiry into Motivation.*

36. Corbin Fowler, *Morality for Moderns.*

37. Giambattista Vico, *The Art of Rhetoric* (*Institutiones Oratoriae,* 1711-1741), from the definitive Latin text and notes, Italian commentary and introduction by Giuliano Crifò, translated and edited by Giorgio A. Pinton and Arthur W. Shippee. A volume in **Values in Italian Philosophy.**

38. W. H. Werkmeister, *Martin Heidegger on the Way,* edited by Richard T. Hull. A volume in **Werkmeister Studies.**

39. Phillip Stambovsky, *Myth and the Limits of Reason.*

40. Samantha Brennan, Tracy Isaacs, and Michael Milde, Editors, *A Question of Values: New Canadian Perspectives in Ethics and Political Philosophy.*

41. Peter A. Redpath, *Cartesian Nightmare: An Introduction to Transcendental Sophistry.* A volume in **Studies in the History of Western Philosophy.**

42. Clark Butler, *History as the Story of Freedom: Philosophy in Intercultural Context,* with Responses by sixteen scholars.

43. Dennis Rohatyn, *Philosophy History Sophistry.*

44. Leon Shaskolsky Sheleff, *Social Cohesion and Legal Coercion: A Critique of Weber, Durkheim, and Marx.*

45. Alan Soble, Editor, *Sex, Love, and Friendship: Studies of the Society for the Philosophy of Sex and Love, 1977-1992.* A volume in **Histories and Addresses of Philosophical Societies.**

46. Peter A. Redpath, *Wisdom's Odyssey: From Philosophy to Transcendental Sophistry.* A volume in **Studies in the History of Western Philosophy.**

47. Albert A. Anderson, *Universal Justice: A Dialectical Approach.* A volume in **Universal Justice.**

48. Pio Colonnello, *The Philosophy of José Gaos.* Translated from Italian by Peter Cocozzella. Edited by Myra Moss. Introduction by Giovanni Gullace. A volume in **Values in Italian Philosophy.**

49. Laura Duhan Kaplan and Laurence F. Bove, Editors, *Philosophical Perspectives on Power and Domination: Theories and Practices.* A volume in **Philosophy of Peace.**

50. Gregory F. Mellema, *Collective Responsibility.*

51. Josef Seifert, *What Is Life? The Originality, Irreducibility, and Value of Life.* A volume in **Central-European Value Studies.**

52. William Gerber, *Anatomy of What We Value Most.*

53. Armando Molina, *Our Ways: Values and Character,* edited by Rem B. Edwards. A volume in **Hartman Institute Axiology Studies.**

54. Kathleen J. Wininger, *Nietzsche's Reclamation of Philosophy.* A volume in **Central-European Value Studies.**

55. Thomas Magnell, Editor, *Explorations of Value.*

56. HPP (Hennie) Lötter, *Injustice, Violence, and Peace: The Case of South Africa.* A volume in **Philosophy of Peace.**

57. Lennart Nordenfelt, *Talking About Health: A Philosophical Dialogue.* A volume in **Nordic Value Studies.**